Math Tricks

The Surprising Wonders of Shapes and Numbers

Alfred S. Posamentier

Prometheus Books

Guilford, Connecticut

 Prometheus Books

An imprint of The Rowman & Littlefield Publishing Group, Inc.
4501 Forbes Blvd., Ste. 200
Lanham, MD 20706
www.rowman.com

Distributed by NATIONAL BOOK NETWORK

British Library Cataloguing in Publication Information Available

Library of Congress Cataloging-in-Publication Data
Name: Posamentier, Alfred S., author.
Title: Math tricks : the surprising wonders of shapes and numbers / Alfred S. Posamentier.
Description: Lanham, MD : Prometheus, [2021] | Includes bibliographical references and
    index. | Summary: "In his latest book, mathematician Alfred S. Posamentier provides
    easily understandable, easily presentable and easily replicated tricks that one can do with
    mathematics"—Provided by publisher.
Identifiers: LCCN 2020042779 (print) | LCCN 2020042780 (ebook) |
    ISBN 9781633886643 (cloth) | ISBN 9781633886650 (ebook)
Subjects: LCSH: Mathematical recreations. | Mental arithmetic. | Geometry.
Classification: LCC QA95 .P657 2021 (print) | LCC QA95 (ebook) | DDC 793.74—dc23
LC record available at https://lccn.loc.gov/2020042779
LC ebook record available at https://lccn.loc.gov/2020042780

∞™ The paper used in this publication meets the minimum requirements of American
National Standard for Information Sciences—Permanence of Paper for Printed Library
Materials, ANSI/NISO Z39.48-1992

To my children and grandchildren, whose future is unbounded: Lisa, David, Daniel, Max, Samuel, Jack, and Charles.

To Barbara for her support, patience, and inspiration.

And in memory of my beloved parents Alice and Ernest, whose support was immeasurable.

Contents

Introduction

In recent years people seem to have gravitated toward cute things said about important subjects. It is no secret that most of them avoid mathematics, largely because school teachers sadly have rarely done much to motivate a love of the subject matter. Moreover, in more recent years secondary school teachers have been inclined to "teach to the test," since their professional teaching evaluation is based on their students' performance on various standardized tests. This trend seems to have usurped time that teachers could have used to enrich their instruction and thereby motivate students to enjoy mathematics. For years the author has developed books intended to enchant readers with mathematical topics that mostly either have gone unnoticed or have not been presented at all. What has not been developed until now is a book that provides easily understandable, easily presentable, and easily replicated tricks within the realm of mathematics that one can perform to show another interesting dimension of the subject. All that this text requires is the ability to do arithmetic, understand the very basics of algebra and geometry, and have an open mind for probability. Nothing presented in this book exceeds the first few years of high school mathematics.

The author's hope is that the many mathematical tricks presented here will be easily understood and replicated, so that the reader can be awed and then impress friends and colleagues with the amazing wonders that mathematics holds. Until now many of these tricks have unfortunately been well-kept secrets. With that in mind, this text presents interesting ways that the reader can use these as a personal "bag of tricks" that could make good dinner talk, or even bring some levity to professional meetings. Perhaps the most important part of these "tricks" is presentation, since most people shy away from mathematics. Consequently, we explain not only what each "trick" is and why it is, but also how best to present it to a non–mathematics-oriented audience.

Naturally, much rests on the style of presentation. Therefore, the trickster should spend some time contemplating how and when to best present the tricks most appropriate for a particular audience—this could be a large group or merely one individual.

This book should attract readers of all ages. The key here is to have fun with mathematics and, as a byproduct, perhaps motivate readers to delve further into the subject matter, which could support further investigation. The underlying intention of presenting these mathematics tricks is to enable audiences to recognize and appreciate the power and beauty of mathematics.

1

Arithmetic Tricks

When we begin our education in elementary school, our first experience of mathematics is through arithmetic. We learn arithmetic as an automatic mechanical procedure, which in recent years has been heavily supported by an electronic calculator—either as a separate tool or as a smartphone app. Clearly arithmetic is the basis for our work with numbers. However, through learning the basics of arithmetic, especially when teachers fixate on making all students proficient in the four basic operations, much potential insight into our work with numbers is lost. But it is exactly here, in doing arithmetic, that one can find enjoyment and enlightenment in working with numbers. In this chapter, we showcase lots of shortcuts and insights into arithmetic processes, some of which will be most entertaining and ones by which a trickster can certainly impress his or her audience with marvels all too often hidden from the general populace.

Arithmetic tricks can not only be entertaining, but also can be quite useful. Before we embark on the entertaining aspects, we will show some useful tricks that help with arithmetic calculation but could also be somewhat intriguing. Some of these might even be more efficient than using a calculator. However, with a bit of practice with each one, you could firm up the procedure for ready reference when needed.

We know that multiplying by 10 in the base 10 system requires merely placing a zero at the end of the mutiplicand; and, of course, multiplying by 100 requires that we add two zeros. Taking this one step further, when we multiply by 5, we could multiply by 10 and then divide the resulting number by 2. Or in some cases it might be easier to first divide by 2 and then multiply by 10. This is rather simple and usually taught in elementary school. We can take this a step further by considering some methods to multiply numbers by divisors of powers of 10. This is where a trickster can begin to impress the audience.

THE TRICK OF MULTIPLICATION BY DIVISORS
OF POWERS OF 10

Multiplying by factors (or divisors) of powers of 10 is more involved than merely multiplying by powers of 10. In many cases, however, it can also be done mentally. Let's begin by considering multiplication by 25 (a factor of 100). Perhaps we should begin with an example. Let's consider 16×25. Since $25 = \dfrac{100}{4}$, we can write this as $16 \times 25 = 16 \times \dfrac{100}{4} = \dfrac{16}{4} \times 100 = 4 \times 100 = 400$. Rather simple! All we did was divide the multiplicand by four and then multiply by 100, which means adding two zeros to the number. However, more practice can be helpful, so here are some additional examples:

38×25 can be written as $38 \times \dfrac{100}{4} = \dfrac{38}{4} \times 100 = \dfrac{19}{2} \times 100 = 9.5 \times 100 = 950$.

1.7×25 can be written as $\dfrac{17}{10} \times \dfrac{100}{4} = \dfrac{17}{4} \times \dfrac{100}{10} = 4.25 \times 10 = 42.5$.

(In the previous example, we converted as follows: $\dfrac{17}{4} = \dfrac{16}{4} + \dfrac{1}{4} = 4 + 0.25 = 4.25$.)

Some of these tricks are more entertaining than useful. But they do give interesting insights into why we learned various arithmetic procedures in school.

Analogously, we can multiply by 125, since $125 = \dfrac{1000}{8}$. Here are some examples of multiplication by 125—mentally!

$$32 \times 125 = 32 \times \dfrac{1000}{8} = \dfrac{32}{8} \times 1000 = 4 \times 1000 = 4000$$

$$78 \times 125 = 78 \times \dfrac{1000}{8} = \dfrac{78}{8} \times 1000 = \dfrac{39}{4} \times 1000 = 9.75 \times 1000 = 9750$$

$$3.4 \times 125 = 3.4 \times \dfrac{1000}{8} = \dfrac{3.4}{8} \times 1000 = \dfrac{1.7}{4} \times 1000 = 0.425 \times 1000 = 425$$

When multiplying by 50, you can use $50 = \dfrac{100}{2}$, or when multiplying by 20, you can use $20 = \dfrac{100}{5}$. Practice with these special numbers will clearly be helpful to the audience, since many calculations can be done faster this way than on your calculator!

A TRICK TO SQUARE A MULTIPLE OF 5

You might want to challenge your audience to square 85 mentally. The initial reaction is usually "Are you kidding?" Well, here we show you a way you can perform this trick rather easily. Let's consider a general case of a two-digit number $(10a + 5)$, which is a multiple of 5. When we square this number, we get $(10a + 5)^2 = 100a^2 + 100a + 25 = a(a + 1)(100) + 25$. This tells us to take the tens digit, multiply it by the tens digit +1, and add 25. We can do this with the following example, where we find the value of 85^2. Following this technique, we multiply the tens digit, which is 8, by $8 + 1 = 9$ and then multiply that by 100 and add 25: $85^2 = (8)(9)(100) + 25 = 7225$. With a little practice this can go quite quickly, and the trickster will have shown another useful technique.

A TRICK FOR MULTIPLYING BY 11

Here is a trick that will truly give the trickster a great reputation. Multiplying by 11 mentally can be significantly simpler than using a calculator, yet this trick is hardly ever presented in school. What a shame! The process is as fast as you can write the number down. For example, suppose you would like to multiply 23×11. All you need to do is add the digits, $2 + 3 = 5$, and place that 5 between the two digits in 23 to get 253. What could be easier! We can state this trick as follows:

> To multiply a two-digit number by 11, just add the two digits and place this sum between the two digits.

Just for practice, let's try using this technique when we want to multiply 45 by 11. According to the rule, we add 4 and 5 and place the result between the 4 and 5 to get 495. This trick can get a bit more difficult when the sum of the two digits you need to add results in a two-digit number. What do we do then? We no longer have a single digit to place between the two original digits. So, if the sum of the two digits is greater than 9, we then place the units digit of that sum between the two digits of the number being multiplied by 11 and "carry" the tens digit to be added to the hundreds digit of the multiplicand. Let's try it with 78×11. We find that the sum of the two digits is $7 + 8 = 15$. Therefore, we place the 5 between the 7 and 8, and add the 1 to the 7, to get $[7 + 1][5][8]$ or 858.

You may legitimately ask whether the rule also holds when 11 is multiplied by a number of more than two digits. Let's try a larger number such as 12,345 and use our trick to multiply it by 11.

Here we begin at the right-side digit and add every pair of digits going from right to left.

$$1[1 + 2][2 + 3][3 + 4][4 + 5]5 = 135,795.$$

Suppose we now combine the skills we have garnered through this trick of multiplying by 11 by applying it to a number that requires the more complicated version, where the sum of adjacent digits exceeds 9. Remember, if the sum of two digits is greater than 9, we place the units digit of this two-digit sum in the middle and carry the tens digit to the next place. To enable the trickster to become an expert at this procedure, we will do one of the more complicated versions here: multiplying 56,789 by 11. This process is tedious, and perhaps uncommon, but we show it here to demonstrate the extension of this multiplication trick. Follow along as we do it step-by-step in Figure 1.1.

5[5+6][6+7][7+8][8+9]9	Add each pair of digits between the end digits
5[5+6][6+7][7+8][17]9	Add 8+9 = 17
5[5+6][6+7][7+8+*1*][7]9	Carry the 1 (from the 17) to the next sum
5[5+6][6+7][16][7]9	Add 7+8+ 1 = 16
5[5+6][6+7+*1*][6][7]9	Carry the 1 (from the 16) to the next sum
5[5+6][14][6][7]9	Add 6+7+1 = 14
5[5+6+*1*][4][6][7]9	Carry the 1 (from the 14) to the next sum
5[12][4][6][7]9	Add 5+6+1 =12
5+1[2][4][6][7]9	Add 1+5=6 to get the answer: 624,679

Figure 1.1

This trick for multiplying by 11 could be a highlight in a trickster's repertoire. The audience will not only be impressed with the trickster's cleverness, but may also appreciate knowing this shortcut.

A TRICK TO DETERMINE WHEN A NUMBER IS DIVISIBLE BY 11

The trickster can also be fortified with another novelty considering the number 11, which owes its special status to being 1 greater than the base of our number system, 10. Let's look at the reverse of the previous arithmetic trick, that is, division by 11. The need to determine if a number is divisible by 11 can come up at the oddest times. If you have a calculator at hand, the problem is easily solved. But that is not always the case. Besides, there is a clever "rule" for testing for divisibility by 11 that is worth knowing just for its charm. Besides, it once again shows the trickster's "cleverness"!

The rule is quite simple:

If the difference of the sums of the alternate digits is divisible by 11, then the original number is also divisible by 11.

This sounds more complicated than it is. Let us take this rule a piece at a time. The phrase "sums of the alternate digits" means that you begin at one end of the number, taking the first, third, fifth (etc.) digits and adding them. Then you add the remaining (second, fourth, etc.) digits. Subtract the two sums and inspect for divisibility by 11. If the resulting number is divisible by 11, then the original number was divisible by 11. And the reverse is also true. That is, if the number reached by subtracting the two sums is *not* divisible by 11, then the original number also was *not* divisible by 11.

This is probably best demonstrated through an example. Suppose we test 768,614 for divisibility by 11. Sums of the alternate digits are: $7 + 8 + 1 = 16$, and $6 + 6 + 4 = 16$. The difference of these two sums, $16 - 16 = 0$, is divisible by 11. (Remember $\frac{0}{11} = 0$.) Therefore, we can conclude that 768,614 is divisible by 11.

Another example might help firm up your understanding of this procedure. To determine if 918,082 is divisible by 11, we need to find the sums of the alternate digits:

$$9 + 8 + 8 = 25, \text{ and } 1 + 0 + 2 = 3.$$

Their difference is $25 - 3 = 22$, which is divisible by 11, and so 918,082 is divisible by 11. Now practice with this rule so that you can demonstrate it more easily and impress your audience by stressing the power and consistency of mathematics.

In case you're asked why this trick works, we offer a brief discussion to justify this rule. Consider the number *ab,cde*, whose value can be expressed as

$$N = 10^4 a + 10^3 b + 10^2 c + 10d + e = (11 - 1)^4 a +$$
$$(11 - 1)^3 b + (11 - 1)^2 c + (11 - 1)d + e$$
$$= [11M + (-1)^4]a + [11M + (-1)^3]b + [11M + (-1)^2]c +$$
$$[11 + (-1)]d + e$$
$$= 11M[a + b + c + d] + a - b + c - d + e,$$

which implies that since the first part of this value of N, $11M[a + b + c + d]$, is already a multiple of 11, the divisibility by 11 of N depends on the divisibility of the remaining part, $a - b + c - d + e = (a + c + e) - (b + d)$, which is actually the difference of the sums of the alternate digits. (Note: $11M$ refers to a multiple of 11.)

THE TRICK FOR DETERMINING WHEN A NUMBER IS DIVISIBLE BY 3 OR 9

Sometimes in everyday life there are situations where it is useful to know whether a number can be divisible by 3 or by 9, especially if the calculation can be mentally done instantly. The trickster will have a fine opportunity to impress the audience with this technique. For example, suppose you are in a restaurant and receive a check of $71.23 , and you want to add a tip, but you want the amount to be split into three equal parts. Wouldn't it be nice if there were some mental arithmetic trick for determining this? Well, here comes mathematics to the rescue. Here we provide you with a simple technique to determine if a number is divisible by 3 and (as an extra bonus) also divisible by 9.

The rule, simply stated, is:

If the sum of the digits of a number is divisible by 3 (or 9), then the original number is divisible by 3 (or 9).

As before, an example may firm up understanding of this rule. Consider 296,357. Let's test it for divisibility by 3 (or 9). The sum of the digits is $2 + 9 + 6 + 3 + 5 + 7 = 32$, which is not divisible by 3 or 9. Therefore, the original number is divisible neither by 3 nor 9.

Let's now assume a group of three people is given a restaurant check of $71.23 and would like to give an approximately 20% tip. They decide to add $14 to the bill, which would make the total $85.23 . They would like to divide the check equally among the three guests. The procedure would have them add the digits to get $8 + 5 + 2 + 3 = 18$, which is divisible by 3. Therefore, the check can be equally divided amongst the three customers. If the number obtained from the sum of the digits is not easily identifiable as a multiple of 3, then continue to add the digits of that number until you reach a number that you can recognize as a multiple of 3. In this case, note that the final result, 18, is also divisible by 9, which implies that the original number was divisible by 9 as well.

Once again, the trickster might be asked why this trick works. So, we now briefly discuss how this rule works. Consider the number ab,cde, whose value can be expressed as follows (where $9M$ refers to a multiple of 9):

$$N = 10^4 a + 10^3 b + 10^2 c + 10d + e = (9+1)^4 a + (9+1)^3 b +$$
$$(9+1)^2 c + (9+1)d + e$$
$$= [9M + (1)^4]a + [9M + (1)^3]b + [9M + (1)^2]c + [9 + (1)]d + e$$
$$= 9M[a + b + c + d] + a + b + c + d + e,$$

which implies that divisibility by 3 or 9 of the number N depends on the divisibility of $a + b + c + d + e$, which is the sum of the digits.

Another example: Is 457,875 divisible by 3 or 9? The sum of the digits is $4 + 5 + 7 + 8 + 7 + 5 = 36$, which is divisible by 9 (and then, of course, by 3 as well), so 457,875 is divisible by 3 and by 9.

As a last example: Is 27,987 divisible by 3 or 9? The sum of the digits is $2+7+9+8+7 = 33$, which is divisible by 3 but not by 9. Therefore, 27,987 is divisible by 3 and not by 9.

Now that you are expert at determining if a number is divisible by 3 or 9, we can go back to our original question about the divisibility of the restaurant bill of $\$71.22 + 14.00 = \$85.22$. Can this amount be divided into three equal parts? Because $8 + 5 + 2 + 2 = 17$, and 17 is not divisible by 3, $\$85.22$ is not divisible by 3, so they will have to modify the tip to have everyone pay the same amount.

A TRICK TO GET THE NUMBER 9

Now that we have some facility with and appreciation for the number 9, let's use this skill in the following trick. Begin by telling your friends that you will generate 9 from various numbers that they provide you. Ask one of them to select any number and add their age to it. Then, have them add to this result the last two digits of their telephone number and multiply the resulting number by 18. Then obtain the sum of the digits of this last number and continue to take the digit sum of each resulting number until a single digit remains. You will impress your friends by showing them that this last digit is 9.

To practice this technique, let's try it with 39 and add our age (37) to get 76. We then add the last two digits of our telephone number (31) to get 107 and multiply this number by 18 to get 1926. The digit sum is $1 + 9 + 2 + 6 = 18$, whose digit sum in turn is $1 + 8 = 9$. This works because when we multiplied by 18, we made sure that the final result would be a multiple of 9, which eventually always yields a digit sum of 9.

THE TRICK TO DETERMINE DIVISIBILITY BY PRIME NUMBERS

In a previous section we presented a nifty little trick for determining if a number is divisible by 3 or by 9. Most adults can determine whether a number is divisible by 2 or by 5, simply by looking at the last digit (i.e., the units digit) of the number. If the last digit is an even number (such as 2, 4, 6, 8, 0), then the number will be divisible by 2. Incidentally, you can extend this simple well-known trick to determine divisibility by higher powers of 2, such as by 4, 8, 16, and so on. Once again, we look at the end of the number being inspected. If the number formed by the last two digits is divisible by 4, then the entire number itself is divisible by 4. Also, if the number formed by the last three digits is divisible by 8, then the number itself is divisible by 8. And so, to test for divisibility by 16 we would focus on the last four digits of the number, and so on for other higher powers of 2.

Analogously, we can develop a trick for determining divisibility by powers of 5. If the last digit of the number being inspected for divisibility is either a 0 or 5, then the number itself will be divisible by 5. If the number formed by the last two digits is divisible by 25, then the number itself is divisible by 25. This is analogous to the trick we used to determine divisibility by powers of 2. Have you guessed what the relationship here is between 2 and 5? Yes, they are the factors of 10, the basis of our decimal number system. Are there also tricks for divisibility by other numbers? What about prime numbers?

With the calculator there is no longer a practical need to know the numbers by which a given number is divisible. You can simply do the division on a calculator. Yet, for a better appreciation of mathematics, and to impress others with a trick of this sort, divisibility rules provide an interesting "window" into the nature of numbers and their properties. For this reason (among others), the topic of divisibility still has a place on the mathematics-learning spectrum and can entertain and, of course, impress friends with these tricks.

The most perplexing tricks have always been those to determine divisibility by prime numbers. This is especially true for divisibility by 7, which follows a series of nifty tricks for 2 through 6. By the way, to determine divisibility by 6 we simply apply the technique for divisibility by 2 and by 3—both must hold true for a number to be divisible by 6. As you will soon see, some of the divisibility tests for prime numbers are almost as cumbersome as actual division algorithms. Yet, they are fun and, believe it or not, can be used to impress others. These tricks often stir up motivation to determine why they work. So now we shall consider divisibility tricks for prime numbers.

We begin by considering how to determine if a given number is divisible by 7 and then, as we inspect the trick, how it can be generalized for other prime numbers.

The trick for divisibility by 7 is as follows:

Delete the last digit from the given number, and then subtract twice this deleted digit from the remaining number. If the result is divisible by 7, the original number is divisible by 7. This process may be repeated until the result can be determined by simple inspection for divisibility by 7.

Let's try an example of how this rule works. Suppose we want to test 876,547 for divisibility by 7. Begin with 876,547 and delete its units digit, 7, and subtract its double, 14, from the remaining number: $87,654 - 14 = 87,640$. Since we cannot yet visually inspect the resulting number for divisibility by 7, we continue the process with the resulting number 87,640. By deleting its units digit, 0, and subtracting its double, still 0, from the remaining number we get $8764 - 0 = 8764$. This did not bring us any closer to visually being able to check for divisibility by 7; therefore, we continue the process with the resulting number 8764. By deleting its units digit, 4, and subtracting its double, 8, from the remaining number we get $876 - 8 = 868$. Since we still cannot visually inspect the resulting number, 868, for divisibility by 7, we continue the process with the resulting number 868. By deleting its units digit, 8, and subtracting its double, 16, from the remaining number we get $86 - 16 = 70$, which we can easily see is divisible by 7. Therefore, the original number 876,547 is also divisible by 7.

Before we continue with our discussion of divisibility of prime numbers, take a few minutes to practice this trick with a few randomly selected numbers and then check your results with a calculator.

Now for the beauty of mathematics! Why does this rather strange procedure actually work? The wonderful thing about mathematics is that it generally doesn't do things that we cannot justify. This procedure will make complete sense to you after you see what is happening in it. By the way, some phenomena in mathematics have not yet been provided an acceptable justification (or proof), but that doesn't mean one won't be found in the future. It took 358 years to justify Pierre de Fermat conjecture that there are no positive integers a, b, and c, for the equation $a^n + b^n = c^n$ for integers $n > 2$. It was done by Dr. Andrew Wiles in 1994.

To justify the trick of determining divisibility by 7, consider the various possible terminal digits that you are "dropping" and the corresponding subtraction done by dropping the last digit. The chart (Figure 1.2) shows you how dropping the terminal digit, doubling it, and subtracting the resulting number gives us in each case a multiple of 7. That is, we have taken "bundles of 7" away from the original number. Therefore, if the remaining number is divisible by 7, then so is the original number. Because you have separated the original number into two parts, each of which is divisible by 7, the entire number must be divisible by 7.

Terminal digit	Number subtracted from original	Terminal digit	Number subtracted from original
1	$20 + 1 = 21 = 3 \times 7$	5	$100 + 5 = 105 = 157$
2	$40 + 2 = 42 = 6 \times 7$	6	$120 + 6 = 126 = 18 \times 7$
3	$60 + 3 = 63 = 9 \times 7$	7	$140 + 7 = 147 = 21 \times 7$
4	$80 + 4 = 84 = 12 \times 7$	8	$160 + 8 = 168 = 24 \times 7$
		9	$180 + 9 = 189 = 27 \times 7$

Figure 1.2

The trick for divisibility by 13 is as follows:

This is similar to the rule for testing divisibility by 7, except that the 7 is replaced by 13, and instead of subtracting twice the deleted digit, we subtract nine times the deleted digit each time.

Let's try our trick to check 5616 for divisibility by 13. We begin by deleting its units digit, 6, and subtracting $9 \times 6 = 54$ from the remaining number: $561 - 54 = 507$. Since we still cannot visually inspect the resulting number for divisibility by 13, we continue the process with the resulting number 507. We delete its units digit and subtract 9 times this digit ($9 \times 7 = 63$) from the remaining number: $50 - 63 = -13$, which is divisible by 13. Therefore, the original number also is divisible by 13.

Now we might want to see how we determined the "multiplier" 9 in our trick. We sought the smallest multiple of 13 that ends in a 1. That was 91, where the tens digit is 9 times the units digit. Once again, consider the various possible terminal digits and the corresponding subtractions in Figure 1.3.

Terminal digit	Number subtracted from original	Terminal digit	Number subtracted from original
1	$90 + 1 = \ 91 = \ \ 7{\times}13$	5	$450 + 5 = 455 = 35{\times}13$
2	$180 + 2 = 182 = 14{\times}13$	6	$540 + 6 = 546 = 42{\times}13$
3	$270 + 3 = 273 = 21{\times}13$	7	$630 + 7 = 637 = 49{\times}13$
4	$360 + 4 = 364 = 28{\times}13$	8	$720 + 8 = 728 = 56{\times}13$
		9	$810 + 9 = 819 = 63{\times}13$

Figure 1.3

In each case a multiple of 13 is being subtracted one or more times from the original number. Hence, if the remaining number is divisible by 13, then the original number is divisible by 13.

The trick for divisibility by 17 is as follows:

Delete the units digit and subtract 5 times the deleted digit each time from the remaining number until you reach a number small enough to visually determine its divisibility by 17.

We justify the trick for divisibility by 17 as we did those for 7 and 13. Each step of the process requires us to subtract a "bunch of 17s" from the original number until we reduce the number to one that we can visually inspect for divisibility by 17. This time we see that the multiplier is 5, since we will be deducting multiples of 17, such as 51, 102, 153, and so on, from the original number.

The patterns developed in the preceding three divisibility tricks (for 7, 13, and 17) should enable you to produce analogous tricks for testing divisibility by larger primes. Figure 1.4 presents the "multipliers" of the deleted digits for various primes.

To test divisibility by	7	11	13	17	19	23	29	31	37	41	43	47
Multiplier	2	1	9	5	17	16	26	3	11	4	30	14

Figure 1.4

You may want to extend this chart. It's fun, and it will increase your perception of mathematics, while at the same time extending your bag of tricks. You may also want to extend your knowledge of divisibility rules to include composite (i.e., nonprime) numbers. Knowing why the rule given below refers to relatively prime factors and not just any factors is something that will sharpen your understanding of number properties. The reason is that relatively prime factors have independent divisibility rules, whereas other factors may not.

In case your audience is interested in your bag of divisibility tricks, you may want to be able to produce techniques for determining divisibility by composite numbers. Therefore, we offer the following trick: *A given number is divisible by a composite number if it is divisible by each of its relatively prime factors.* (Two numbers are relatively prime if they have no common factors other than 1.) Figure 1.5 offers illustrations of this rule.

To be divisible by	6	10	12	15	18	21	24	26	28
The number must be divisible by	2, 3	2, 5	3, 4	3, 5	2, 9	3, 7	3, 8	2, 13	4, 7

Figure 1.5

At this juncture you have not only an extensive list of tricks for testing divisibility, but also an interesting insight into elementary number theory. Practice using these rules (to instill greater familiarity) and try to develop tricks to test divisibility by other numbers in base 10 and then generalize these tricks to numbers in other bases. Unfortunately, lack of space prevents a more detailed development here. Yet, we hope we have now whetted your appetite regarding divisibility.

A CURIOUS DIVISIBILITY TRICK

With this trick you can entertain an audience even over a dinner table. This trick enables the trickster to easily determine divisibility by numbers from 2 to 9. Perhaps it is best demonstrated through a variety of examples. Suppose we want to determine if 96 is divisible by 8. We simply separate the digits, multiply the tens digit by 2, and add the result to the units digit. So for 96 we do the following: $9 \times 2 = 18 + 6 = 24$. Since 24 is divisible by 8, we can conclude that 96 is also divisible by 8. Now the question is, why did we multiply 9×2? The 2 was selected because it is the difference between 10 and 8. The 6 we added was the original units digit. To firm up this trick, let us do a few others.

This time, we determine whether 182 is divisible by 7. To do that we separate the units digit from the remainder of the number. This time, we will multiply the units digit by 3, since $10 - 7 = 3$. We now do the following: $18 \times 3 = 54 + 2 = 56$, which is divisible by 7, and therefore, we can conclude that 182 also is divisible by 7. The 2 we then added came from the original units digit. Just to be on the safe side, let's consider one further example: determine if 687 is divisible by 7. Remember, our multiplier will once again be 3 (since $10 - 7 = 3$), so our procedure is as follows: $68 \times 3 = 204 + 7 = 211$. Since we cannot mentally immediately determine if 211 is divisible by 7, we simply repeat the procedure. So $21 \times 3 = 63 + 1 = 64$, which is not divisible by 7, and therefore 687 also is not divisible by 7. An ambitious reader may wish to use simple algebra to determine why this trick works.

SIMPLY, A CUTE TRICK

Here is a cute little trick that one can play across the dinner table, since it can be done mentally very easily. Have your audience select a number from 1 to 10 and keep it secret. Then ask them to multiply their secret number by 9. Then ask them to add the digits of the number they arrived at and subtract 5 from that number. Still without divulging any information, have them select the letter of the alphabet that corresponds with the number they arrived at ($A = 1, B = 2, C = 3$, etc.). Have them select a country in Europe, whose first letter is the same as the one that they have identified. Then ask them to select an animal whose name begins with the last letter of the European country they identified. You, as the trickster, will be able to tell them that the animal that they have identified is a kangaroo. They will surely be amazed at your talent as a trickster.

Here is the reason for your hidden "talent." When they add the digits of the number that they obtained by multiplying the original number by 9, the result will have to be a 9, as we have already discussed earlier in this chapter. Subtracting 5 from the 9 yields 4. The fourth letter of the alphabet is a D. The only country in Europe whose name begins with a D is Denmark. The only *popular* animal whose name begins with the last letter of the word Denmark (K) is the kangaroo. And so, you have used a few trick talents to impress your audience.

AN ENTERTAINING TRICK

This is another trick that you can use to simply entertain your audience. Tell them that you can select a six-digit number and do some manipulation to get back to the same number you started with. The number you would select is 998,001. The process will be as follows. You should make the process clear, since otherwise it will sound rather strange to the audience. So tell them that you will separate it into two parts, each of three digits, and then add these two three-digit numbers, and square the sum. The result will be the original six-digit number. After the audience has grasped this process, you can do it with your previously selected number 998,001. When we split the number in half and square it, we get: $(998 + 001)^2 = (999)^2 = 998,001$. If your audience needs further evidence that you have "trickster talent," you can repeat the process with 494,209, since $(494 + 209)^2 = (703)^2 = 494,209$. Ambitious audiences might wish to select other such examples.

GUESSING THE MISSING NUMBER

Guessing a number always catches the audience's curiosity. Begin by presenting the three numbers 1, 3, and 8 and ask the audience for a fourth number, n, such that when the product of *any pair* of the four numbers is added to 1, the result will be a square number. In other words, if we take the product of two of the given numbers, say, $3 \times 8 = 24$ and add it to 1, we get 25, which is a square number. Another possibility would be $8 \times 1 = 8$, which, when added to 1, gives 9, which also is a square number. Typically, the audience starts trying to replace the number n with smaller numbers and finds that their selected numbers don't hold this pattern for every pair from the four numbers being considered.

Here we provide the answer: $n = 120$, since 1×120, which, when added to 1, is 11^2; and $3 \times 120 = 360$, which, when added to 1, is $361 = 19^2$; and $8 \times 120 = 960$, which when added to $1 = 961 = 19^2$. Although this may take some time on the audience's part, this could be entertaining, since the sought-after number is quite far from the other given numbers.

GUESSING THE MULTIPLIER

Certain numbers have properties that enable the trickster to impress the audience. Here is once such trick you can play on your friends to show how clever you are. Ask your audience to take any of the following numbers and multiply it by 37:

3, 6, 9, 12, 15, 18, 21, 24, 27.

After they tell you what the product is, you will be able to tell them *immediately* which number they used as a multiplier. The reason is that each of the potential multipliers will result in a three-digit number, whose digits all are the same. And the sum of those digits will give you the sought-after multiplier.

For example, supposing your audience multiplied $37 \times 21 = 777$. The sum of these digits $7 + 7 + 7 = 21$. Let's take another example to see if you can discover the trick. Suppose we multiply $37 \times 15 = 555$, or perhaps $37 \times 6 = 222$. By now you should notice that the resulting three-digit number will be the one whose digit sum is the sought-after multiplier from the given list. The impressive thing about this trick is that you will be able to provide the number as quickly as they give you the product they obtained.

THE PRODUCT OF TWO NEGATIVE NUMBERS

Sometimes a "math person" is asked to explain a common concept that is often accepted without much explanation. The mathematics trickster could well be asked to explain why the product of two negative numbers is a positive number. We know that logically, when two negative statements are made in one sentence, the result is positive. For example, the sentence "She does not take no money out of her bank account" in effect translates to mean "She does take money from her bank account." Mathematics often asks the question, how can we demonstrate that the product of two negatives is a positive—say, algebraically. Here is a trick for doing this using the distributive property.

We begin with an equality to which 0 is added, leaving the equality intact.

$$(-a)(-b) = (-a)(-b) + (0)(b)$$

Then substituting $0 = (a - a)$:

$$(-a)(-b) = (-a)(-b) + (-a + a)(b)$$

Applying the distributive property, we get:

$$(-a)(-b) = (-a)(-b) + (-a)(b) + (a)(b)$$

Once again, the distributive property yields:

$$(-a)(-b) = (-a)(-b + b) + (a)(b)$$

The rest should be obvious!

$$(-a)(-b) = (-a)(0) + (a)(b)$$
$$(-a)(-b) = (a)(b)$$

There are many other ways to show that the product of two negatives is a positive. For example, if you don't take apples out of a basket of apples, then you will have apples in the basket. The trickster may also decide to demonstrate this property by showing a progression of multiplications:

$$-3 \times 2 = -6$$
$$-3 \times 1 = -3$$
$$-3 \times 0 = 0$$
$$-3 \times -1 = +3$$
$$-3 \times -2 = +6, \text{ and so on.}$$

However, the previous procedure might be considered more mathematically elegant and convincing. Here we let the reader choose.

SOME SHORT TRICK QUESTIONS

Sometimes a trickster has to squeeze a trick into a very short time span and still make a big splash. We offer a few such "quickies" here.

1. Which two whole numbers when multiplied yield 197?
   The answer is quite simple: 1 and 197.
2. How can you write 100 using exactly 10 digits?
   The answer is $99\dfrac{9999}{9999}$
3. The trick is to find two numbers, each of which has the sum of its digits equal to 45 and the difference of the numbers has a sum of digits also equal to 45. The answer is:

$$987,654,321 - 123,456,789 = 864,197,532.$$

4. Of the three numbers provided here, which digits can you erase so that the sum of the remaining numbers is equal to 20?

$$111$$
$$777$$
$$999$$

The answer is:

$$\cancel{+}11$$
$$\cancel{777}$$
$$\cancel{999}$$

This will leave you with $11 + 9 = 20$.

5. Find the fraction that, when reduced by $\frac{1}{7}$ of itself, is equal to $\frac{1}{8}$.

   The answer is $\frac{7}{48}$, since $\frac{1}{7}$ of it, $\frac{1}{48}$, subtracted from $\frac{7}{48}$ yields $\frac{1}{8}$, as

   shown here: $\frac{7}{48} - \frac{1}{48} = \frac{6}{48} = \frac{1}{8}$.

6. Create 10 using all 10 digits.

   Here we create 10 by using all digits from 0 to 9.

   One answer is: $1\frac{35}{70} + 8\frac{46}{92} = 10$

7. Create 100 using all 10 digits.

   Here we are asked to create 100 by using all digits from 0 to 9.

   The answer is: $(9 \times 8) + 7 + 6 + 5 + 4 + 3 + 2 + 1 + 0$.

8. Find the number that is 11 times a sum of its digits. (It may be helpful to look back to the trick on multiplying by 11.) A clever trickster should see that the required number is 198, since $198 = 11 \times (1 + 9 + 8)$.

SOME PRIME DENOMINATOR SURPRISES

Recall that prime numbers have no factors other than themselves and 1. We now will consider the fractions whose denominators are prime numbers (excluding 2 and 5), and where the decimal expansion yields an *even* number of repeated digits. This provides the trickster with some highly unusual relationships that will surely impress the audience. We offer here a few fractions with an even number of repeated digits:

$$\frac{1}{7} = .142857142857142857... = .\overline{142857}$$

$$\frac{1}{11} = .090909... = .\overline{09}$$

$$\frac{1}{13} = 076923076923... = .\overline{076923}$$

$$\frac{1}{17} = .\overline{0588235294117647}$$

$$\frac{1}{19} = .\overline{052631578947368421}$$

$$\frac{1}{23} = 0.\overline{0434782608695652173913}$$

We will now show a most unanticipated trick move. We will treat each of these repeating periods as a new number, split its even-numbered sequence of digits into two equal parts, and then add them. The trickster can now amaze the audience with surprising results when we add the split parts of the repetitive sequence,[1] as shown in Figure 1.6.

$\frac{1}{7} = .142857142857142857K = .\overline{142857}$	142 857 999
$\frac{1}{11} = .090909K = .\overline{09}$	0 9 9
$\frac{1}{13} = .\overline{076923}$	076 923 999
$\frac{1}{17} = .\overline{0588235294117647}$	05882352 94117647 99999999
$\frac{1}{19} = .\overline{052631578947368421}$	052631578 947368421 999999999
$\frac{1}{23} = 0.\overline{0434782608695652173913}$	04347826086 95652173913 99999999999

Figure 1.6

The awe that this result engenders may entice the audience to find further examples—another generation of wonder. The results of these additions doubtless will truly impress your audience.

A FAVORITE DIGIT TRICK

You might try this little calculating "trick" with your friend. Write the following number on a piece of paper: 12,345,679 (notice the 8 is missing). Then have your friend multiply 9 by his favorite digit (selected from the digits 1 − 9). Your friend then should multiply this product by the magic number 12,345,679, and to your friend's amazement the product will consist of only his favorite digit.

Favorite digit **4**	*Favorite digit* **2**	*Favorite digit* **5**
Multiplication **4 × 9 = 36**	Multiplication of **2 × 9 = 18**	Multiplication of **5 × 9 = 45**
12,345,679 × 36 =	12,345,679 × 18 =	1,2345,679 × 45 =
444,444,444	222,222,222	555,555,555

Figure 1.7

For example, consider the following three cases shown in Figure 1.7, where your friend may select 4, 2, and 5 as favorite digits.

From this you can see how to have some more fun with arithmetic.

THE 777 NUMBER TRICK

Once again, the nature of numbers enables us to create an impressive trick. Have your friend select any number between 500 and 1000 and then add 777 to this number. If the sum exceeds 999, then have your friend remove the thousands digit and add it to the units digit. Then ask to have the two numbers subtracted—that is, the sum and the number originally selected. You can then astonish your friend by determining the result is 222.

To secure an understanding of the procedure, let's try one example. Suppose your friend selects 600. He or she then adds 777 to it to get $600 + 777 = 1377$. The thousands digit, 1, is then removed and added to the units digit to get $377 + 1 = 378$. Subtracting these two numbers (his original number and the one just obtained), $600 - 378 = \mathbf{222}$.

You may wonder how this trick works. For every selected number between 500 and 1000 you will always get a 1 in the thousands place when the number is added to 777. Dropping the 1 and adding it to the units digit is tantamount to merely subtracting 999 from the number. That is, $-999 = -1000 + 1$. If we now represent the selected number as n, then $n - (n + 777 - 999) = n - n - 777 + 999 = 222$. Remember, n represents the randomly selected number.

Suppose we had used a number other than 777 as our "magic" number, say 591. Then our friend would end up with 408 every time, regardless of which number he chose between 500 and 1000. For a "magic" number of 733, the end result will always be 266. Remember, the selected number cannot be less than 500, or else you will not get a sum in the thousands. At the same time the selected number should not be greater than 999, or you might get a 2 in the thousands place, which also would ruin this scheme.

A SLEIGHT-OF-HAND MATHEMATICS TRICK

Some tricks are somewhat independent of mathematical peculiarities. Rather, they are dependent on a clever form of "manipulation." Here is one such mathematics trick.

Ask each person in your audience to randomly select four five-digit numbers and write them in such a way that they can be added. They should then show them to you. You will then add another set of four selected five-digit numbers. You will then compete with an audience member to add all these eight selected five-digit numbers—even if they use a calculator! It is perhaps best to see how this works to the trickster's advantage.

Let's try this trick with the following example:

Assume that the audience member selected these four five-digit numbers:

$$35,862$$
$$13,579$$
$$27,428$$
$$55,073$$

Your task now, as the trickster, is to augment this list with four additional five-digit numbers:

$$64,137$$
$$86,420$$
$$72,571$$
$$44,926$$

When we add these eight five-digit numbers, the answer will be 399,996, which you will compute much more quickly than your audience member using a calculator!

The question now arises, how did the trickster select the second group of four numbers? We selected each of our four subsequent numbers, digit by digit, so that each of the digits of the first number we selected, when added respectively to the digits of the friend's first number, will add up to 9. Comparing the audience member's first number (35,862) and the trickster's first number (64,137), we notice that each of the respective pairs of digits has a sum of 9:

$$3 + 6 = 9, \ 5 + 4 = 9, \ 8 + 1 = 9, \ 6 + 3 = 9, \ 2 + 7 = 9.$$

We then select our remaining three numbers the same way. This shows that we are actually adding 99,999 four times, which is equal to 399,996.

You could also do this trick with four-digit numbers, where you will end up with $4 \times 9999 = 39{,}996$. If you choose to work with three-digit numbers your expected answer would be $3 \times 999 = 3996$. By now you should see the pattern, so that you can tailor the trick to suit your particular audience. In any case, the audience will be perplexed at how you were able to come up with the sum so quickly. It would not be wise to do this trick a second time unless you use initial numbers with more or fewer digits. You surely do not want to come up with the same sum a second time.

WHEN A QUOTIENT EQUALS A DIFFERENCE

We now offer the trickster the ability to challenge the audience to find two numbers whose quotient and difference are both equal to 5. Typically, the audience will immediately experiment with various integers to see which suit the challenge. Unfortunately, that would be the wrong approach. When the quotient of two numbers is 5, then their difference will be four times the smaller of the two numbers. For example, the quotient of $30 \div 6 = 5$, and the difference between $30 - 6 = 24$, which is 5×6.

One possible pair of numbers that satisfies this relationship is $\dfrac{25}{4}$ and $\dfrac{5}{4}$. The difference of these two numbers is 5, as is their quotient.

In general, when $x - y = \dfrac{x}{y} = a$, we have $x = \dfrac{a^2}{a - 1}$ and $y = \dfrac{a}{a - 1}$.

Another pair of numbers that satisfies this relationship is $\dfrac{144}{11}$ and $\dfrac{12}{11}$. Their difference is $\dfrac{132}{11} = 12$ and their quotient is $\dfrac{144}{11} \div \dfrac{12}{11} = \dfrac{144}{11} \cdot \dfrac{11}{12} = 12$. Can you find other such numbers?

THE TRICK OF SHOWING THAT 5 = 4

This trick will surely have your audience guessing and also a bit uncomfortable until you expose the secret behind it. To do the trick, we are given that $a + b = c$. Certainly it is true that $4a + 4b - 5c = 4a + 4b - 5c$.

When we add these two equations, we get the following: $5a + 5b - 5c = 4a + 4b - 4c$.

By factoring each side of the equation, we get: $5(a + b - c) = 4(a + b - c)$.

When we divide both sides by the common factor $(a + b - c)$, we get $5 = 4$.

Pause to allow your audience to marvel over this ridiculous result. You can then expose the error, namely, that division by zero is forbidden in mathematics. Since we began with the understanding that $a + b = c$, it follows that $(a + b - c) = 0$. Therefore, we divided both sides by 0, which leaves us with a ridiculous result. This justifies why division by zero is forbidden in mathematics.

HOWLERS

In our early years of schooling we learned to reduce fractions to make them more manageable. There were specific ways to do it correctly. Some tricksters seem to have come up with a shorter way to reduce some fractions. Are they correct?

Suppose you are asked to reduce the fraction $\dfrac{26}{65}$, and you do it in the following way:

$$\frac{2\cancel{6}}{\cancel{6}5} = \frac{2}{5}$$

That is, you just cancel out the 6's to get the right answer. Is this procedure correct? Can it be extended to other fractions? If it can, then our elementary school teachers surely treated us unfairly by making us do much more work. Let's look at what was done here and see if it can be generalized. In his book *Fallacies in Mathematics*, the British mathematician E. A. Maxwell refers to the following cancellations as "howlers":

$$\frac{1\cancel{6}}{\cancel{6}4} = \frac{1}{4} \quad \frac{2\cancel{6}}{\cancel{6}5} = \frac{2}{5}$$

Perhaps when someone did fraction reductions this way and still got the right answer, it could just make you howl. As we look at this awkward—yet easy—procedure, we could actually use it to reduce the following fractions to lowest terms:

$$\frac{16}{64}, \frac{19}{95}, \frac{26}{65}, \frac{49}{98}.$$

After reducing each of the fractions to lowest terms in the usual manner, one might ask why it couldn't have been done in the following way:

$$\frac{1\cancel{6}}{\cancel{6}4} = \frac{1}{4}$$

$$\frac{1\!\!\!/9}{9\!\!\!/5} = \frac{1}{5}$$

$$\frac{2\!\!\!/6}{6\!\!\!/5} = \frac{2}{5}$$

$$\frac{4\!\!\!/9}{9\!\!\!/8} = \frac{4}{8} = \frac{1}{2}$$

At this point you may be somewhat amazed. Your first reaction is probably to ask, can this be done to any fraction composed of two-digit numbers of this sort? Can you find another fraction (comprised of two-digit numbers) for which this type of cancellation will work? You might cite $\frac{55}{55} = \frac{5}{5} = 1$ as an illustration of this type of cancellation. Clearly, this will hold true for all two-digit multiples of 11.

For readers with a good working knowledge of elementary algebra, we can "explain" this awkward occurrence. Are the four fractions above the *only* ones (composed of different two-digit numbers) where this type of cancellation holds true?

Consider the fraction $\dfrac{10x + a}{10a + y}$.

The above four cancellations were such that, when the a's were canceled, the fraction was equal to $\dfrac{x}{y}$.

Therefore, $\dfrac{10x + a}{10a + y} = \dfrac{x}{y}$.

This yields: $y(10x + a) = x(10a + y)$

$$10xy + ay = 10ax + xy$$

$$9xy + ay = 10ax$$

And so $y = \dfrac{10ax}{9x + a}$

At this point we shall inspect this equation. It is necessary that x, y, and a be integers since they were digits in the numerator and denominator of a fraction. Our task now is to find the values of a and x for which y will also be integral. To avoid a lot of algebraic manipulation, you need to set up a chart that will generate values of y from the equation $y = \dfrac{10ax}{9x + a}$. Remember, x, y,

x\a	1	2	3	4	5	6	...	9
1		$\dfrac{20}{11}$	$\dfrac{30}{12}$	$\dfrac{40}{13}$	$\dfrac{50}{14}$	$\dfrac{60}{15}=4$		$\dfrac{90}{18}=5$
2	$\dfrac{20}{19}$		$\dfrac{60}{21}$	$\dfrac{80}{22}$	$\dfrac{100}{23}$	$\dfrac{120}{24}=5$		
3	$\dfrac{30}{28}$	$\dfrac{60}{29}$		$\dfrac{120}{31}$	$\dfrac{150}{32}$	$\dfrac{180}{33}$		
4								$\dfrac{360}{45}=8$
⋮								
9								

Figure 1.8

and a must be single-digit integers. Below is a portion of the table you will be constructing. Notice that the cases where $x = a$ are excluded, since $\dfrac{x}{a} = 1$.

The portion of the chart in Figure 1.8 has already generated two of the four integral values of y; that is, when $x = 1$, $a = 6$, and $y = 4$, and when $x = 2$, $a = 6$, and $y=5$. These values yield the fractions $\dfrac{16}{64}$ and $\dfrac{26}{65}$, respectively. The remaining two integral values of y will be obtained when $x = 1$ and $a = 9$, yielding $y = 5$, and when $x = 4$ and $a = 9$, yielding $y = 8$. These yield the fractions $\dfrac{19}{95}$ and $\dfrac{49}{98}$ respectively. This should convince you that there are only four such fractions composed of two-digit numbers.

You may now wonder: are there fractions with numerators and denominators of more than two digits for which this strange type of cancellation holds true? Try this type of cancellation with $\dfrac{499}{998}$. You should find that $\dfrac{4\not9\not9}{\not9\not98} = \dfrac{4}{8} = \dfrac{1}{2}$.

A pattern is now emerging, and you may realize that

$$\frac{4\not9}{\not98} = \frac{4\not9\not9}{\not9\not98} = \frac{4\not9\not9\not9}{\not9\not9\not98} = \frac{4\not9,\not9\not9\not9}{\not9\not9,\not9\not98} = \frac{4}{8} = \frac{1}{2}$$

$$\frac{1\not6}{\not64} = \frac{1\not6\not6}{\not6\not64} = \frac{1\not6\not6\not6}{\not6\not6\not64} = \frac{1\not6,\not6\not6\not6}{\not6\not6,\not6\not64} = \frac{1}{4}$$

$$\frac{1\not9}{\not95} = \frac{1\not9\not9}{\not9\not95} = \frac{1\not9\not9\not9}{\not9\not9\not95} = \frac{1\not9,\not9\not9\not9}{\not9\not9,\not9\not95} = \frac{1}{5}$$

$$\frac{2\cancel{6}}{\cancel{6}5} = \frac{2\cancel{66}}{\cancel{66}5} = \frac{2\cancel{666}}{\cancel{666}5} = \frac{2\cancel{6},\cancel{666}}{\cancel{66},\cancel{66}5} = \frac{2}{5}$$

If the trickster extends the discussion this far, enthusiastic audience members may wish to justify these extensions of the original howlers. Those who, at this point, want to seek out additional fractions that permit this strange cancellation should consider the fractions shown below. They should verify the legitimacy of this strange cancellation, and then set out to discover more such fractions. This will give tricksters even more examples for their bag of tricks—however they may plan to use them.

$$\frac{3\cancel{3}2}{8\cancel{3}0} = \frac{32}{80} = \frac{2}{5}$$

$$\frac{3\cancel{8}5}{8\cancel{8}0} = \frac{35}{80} = \frac{7}{16}$$

$$\frac{1\cancel{3}8}{\cancel{3}45} = \frac{18}{45} = \frac{2}{5}$$

$$\frac{2\cancel{7}5}{7\cancel{7}0} = \frac{25}{70} = \frac{5}{14}$$

$$\frac{1\cancel{6}\cancel{3}}{\cancel{3}2\cancel{6}} = \frac{1}{2}$$

Besides being an algebraic application that you can use to introduce a number of important topics in a motivational way, this topic can also provide recreation. Here are some more of these "howlers."

$$\frac{4\cancel{8}\cancel{4}}{\cancel{8}\cancel{4}7} = \frac{4}{7} \quad \frac{\cancel{3}\cancel{4}5}{6\cancel{3}\cancel{4}} = \frac{5}{6} \quad \frac{\cancel{4}24}{7\cancel{4}\cancel{2}} = \frac{4}{7} \quad \frac{24\cancel{9}}{\cancel{9}96} = \frac{24}{96} = \frac{1}{4}$$

$$\frac{4\cancel{8},\cancel{4}\cancel{8}\cancel{4}}{\cancel{8}\cancel{4},\cancel{8}\cancel{4}7} = \frac{4}{7} \quad \frac{\cancel{3}\cancel{4},\cancel{3}\cancel{4}5}{6\cancel{3},\cancel{4}\cancel{3}\cancel{4}} = \frac{5}{6} \quad \frac{\cancel{4}\cancel{2},\cancel{4}24}{7\cancel{4},\cancel{2}\cancel{4}\cancel{2}} = \frac{4}{7}$$

$$\frac{\cancel{3}2\cancel{4}3}{4\cancel{3}2\cancel{4}} = \frac{3}{4} \quad \frac{\cancel{6}\cancel{4}86}{86\cancel{4}\cancel{8}} = \frac{6}{8} = \frac{3}{4}$$

$$\frac{14,7\cancel{1}\cancel{4}}{7\cancel{1},\cancel{4}68} = \frac{14}{68} = \frac{7}{34} \quad \frac{\cancel{8}7\cancel{8},\cancel{0}\cancel{4}8}{9\cancel{8}7,\cancel{8}\cancel{0}\cancel{4}} = \frac{8}{9}$$

$$\frac{1,\cancel{4}2\cancel{8},\cancel{3}7\cancel{1}}{\cancel{4},2\cancel{8}\cancel{8},7\cancel{1}3} = \frac{1}{3} \quad \frac{2,\cancel{8}\cancel{3}7,\cancel{1}\cancel{4}2}{\cancel{8},\cancel{8}7\cancel{1},\cancel{4}26} = \frac{2}{6} = \frac{1}{3} \quad \frac{3,\cancel{4}6\cancel{1},\cancel{3}\cancel{3}8}{\cancel{4},\cancel{6}\cancel{1}\cancel{3},\cancel{3}84} = \frac{3}{4}$$

$$\frac{7\cancel{6}7,\cancel{1}2\cancel{3},2\cancel{8}7}{876,7\cancel{2}\cancel{2},\cancel{3}2\cancel{8}} = \frac{7}{8} \quad \frac{\cancel{3},2\cancel{4}\cancel{3},2\cancel{4}\cancel{3},2\cancel{4}3}{4,\cancel{3}2\cancel{4},\cancel{3}2\cancel{4},\cancel{3}2\cancel{4}} = \frac{3}{4}$$

$$\frac{1,023,641}{4,102,364} = \frac{1}{4} \qquad \frac{3,243,243}{4,324,324} = \frac{3}{4} \qquad \frac{4,571,428}{5,713,283} = \frac{4}{5}$$

$$\frac{767123287}{876712328} = \frac{7}{8} \qquad \frac{324324324 3}{432432432 4} = \frac{3}{4}$$

$$\frac{1023641}{4102364} = \frac{1}{4} \qquad \frac{3243243}{4324324} = \frac{3}{4} \qquad \frac{4571428}{5714285} = \frac{4}{5}$$

$$\frac{3,384,615}{7,538,461} = \frac{5}{7} \qquad \frac{2,031,282}{8,203,128} = \frac{2}{8} = \frac{1}{4} \qquad \frac{3,116,883}{8,311,688} = \frac{3}{8}$$

$$\frac{6,486,486}{8,648,648} = \frac{6}{8} = \frac{3}{4} \qquad \frac{484,848,484}{848,484,847} = \frac{4}{7}$$

You see now how elementary algebra can be used to investigate a number theory situation, one that is also quite amusing. A clever trickster will find multiple applications for the peculiar math demonstrated in this section.

THE TRICK OF ALWAYS ARRIVING AT THE NUMBER 1089

This trick is best done with an audience of several people so that the end result has greater dramatic effect. The trickster here will have everyone select a three-digit number. By following the trickster's direction the entire audience will end up with the same number, 1089. We shall begin by having the audience select any three-digit number in which the units and hundreds digits are not the same. The trickster will then give them the following step-by-step instructions, while we provide an example for each one.

Choose any three-digit number in which the unit and hundreds digits are not the same.
   We arbitrarily select **825**.
Reverse the digits of the number you have selected.
   We reverse the digits of 825 to get **528**.
Subtract the two numbers (the larger minus the smaller).
   Our difference is $825 - 528 = \mathbf{297}$.
Once again, reverse the digits of this difference.
   Reversing the digits of 297, we get **792**.
Now, add your last two numbers.
   We then add the last two numbers to get $297 + 792 = \mathbf{1089}$.

Each member of the audience should get the same result as ours, 1089, even though the audience members probably started with numbers different from ours. Any who claim they obtained a different result, clearly, have made a calculation error. In any case, your audience will probably be astonished that regardless of the number they initially selected, they all got the same result.

If the original three-digit number had the same units and hundreds digits, we would obtain a zero after the first subtraction; for example, for $n = 373$, $373 - 373 = 0$. This would ruin our model. Before reading on, convince yourself that this scheme will work for other numbers. How does it happen? Is this a "freak property" of this number? Did we do something devious in our calculations?

This illustration of a mathematical oddity depends on the arithmetic operations. We assumed that any number we chose would lead us to 1089. How can we be sure? Well, we could try all possible three-digit numbers to see if it works. That would be tedious and not particularly elegant. A proper investigation of this oddity requires nothing more than some knowledge of elementary algebra. Yet were we to test all possibilities, it would be interesting to determine how many three-digit numbers we would have to apply to this scheme. Remember, we can use only those three-digit numbers whose units and hundreds digits are not the same.

Another hidden beauty in this trick can be seen when considering all possible four-digit numbers that can be generated from this curious number 1089. These are shown in the right-side column of Figure 1.9.

1089	1×1089
2178	2×1089
3267	3×1089
4356	4×1089
5445	5×1089
6534	6×1089
7623	7×1089
8712	8×1089
9801	9×1089

Figure 1.9

Looking at all the numbers that can be obtained using this technique, we notice that each of them is a multiple of 1089, and the first four numbers are reversals of the last four numbers, and the fifth number is a palindrome, that is, a reversal of itself. Also, note that the last entry in Figure 1.9 shows that 9801 is a multiple of its reversal, 1089. You might want to impress your audience by telling them there is only one number of five distinct digits whose multiple is a reversal of the original number. That number is 21,978, since $4 \times 21,978 = 87,912$, its reverse number. Similarly structured numbers are produced with 1089 that provide more material for the trickster, as we will see in the next trick.

THE UNEXPECTEDLY IRREPRESSIBLE NUMBER 10,989 AND OTHERS

The previous trick can be extended to larger numbers. There are many oddities in mathematics that can provide entertainment as well as motivate research. We will consider one here. The reader will be left to discover why this oddity works; however, it will also provide the reader with a cute little "trick" to share with friends. For example, suppose you ask your friends to write any four-digit number on a piece of paper, where the difference between the first and last digits is greater than 1. Then you have them interchange the first and last digits. Next, they are to subtract the smaller from the larger of these two numbers. They then interchange the first and last digits of the subtraction result and add two of the previously obtained numbers. They will get 10,989.

To illustrate how this works, we shall select a random four-digit number in which the units digit and the thousandths digit differ by more than 1. Suppose we select 3798. Following the previously described procedure, we interchange the first and last digits to get 8793. We then subtract these two numbers, $8793 - 3798 = 4995$. Once again, we interchange the first and last digits to get 5994, then add this to the previously obtained number to get $4995 + 5994 = 10,989$, as predicted!

Let's try this trick with the five-digit number 58,632. When we interchange the first and last digits, we get 28,635. We now subtract these two numbers: $58,632 - 28,635 = 29,997$. We then interchange the first and last digits to get 79,992. Now we add the last two numbers: $29,997 + 79,992 = 109,989$. This will always be our final result regardless of which five-digit number we start with.

If you want to further impress your friends, have them select a six-digit number and follow the same procedure. They will always find that the end result is 1,099,989. With a seven-digit number, the result analogously will be 10,999,989. This same pattern then continues for larger numbers. This trick can generate some genuine interest in the nature of numbers and, of course, make mathematics come alive.

THE TRICK TO GUESS THE FOUR-DIGIT NUMBER

This trick will have your friends generate a four-digit number, which you will then determine without them telling you what their number was. Tell them to generate their four-digit number using the following procedure:

1. Select a number from 1 to 9.
2. Place the 0 at the end of the number, forming a two-digit number.
3. Add the original selected digit to this two-digit number.
4. Multiply this two-digit number by 3.
5. Then multiply this product by 11.
6. Finally, multiply the just-obtained number by 3.
7. Ask your friends to divulge the units digit of their answer.

Just by telling you the units digit you will be able to determine the number that they ended up with. In order to guess the four-digit number you will do the following: You subtract from 9 the units digit just divulged to you. This will give you the hundreds digit of the number you ultimately are seeking. The thousands digit will always be one greater than the hundreds digit. To get the tens digit you merely subtract the thousands digit from 9.

To see this process work, we will do it using the initial number 7. The first request was to place a 0 after the 7 to get 70. To that we then add the initial number 7 to get 77. When we multiply by 3 we get 231. Multiplying that by 11 gives us 2541. When we multiply this number by 3, we get 7623. At this point, we divulge the units digit, 3, and the trickster now has to do some work to guess 7623.

The trickster then gets the hundreds digit by subtracting the given units digit 3 from 9 to get 6. The thousands digit is simply one greater than the 6, which is 7. All that remains now is to find the tens digit, which is obtained by subtracting the newly found thousands digit, 7, from 9 to get 2. Hence, the trickster has discovered that the number must be 7623, which is correct!

THE "DETERMINE YOUR DELETED NUMBER" TRICK

The interaction between a trickster and a player might go like that shown in Figure 1.10.

The player tends to try this a few times only to find that the trickster can somehow determine which number has been removed. How is this possible? Yes, some knowledge of number properties will help you figure out this secret.

The trickster says:	The player
Select any five-digit number (not one with five of the same digits) and keep it secret.	35,630
Rearrange the digits of your selected number in any way you wish.	53,306
Subtract these two numbers.	$53,306 - 35,630 = 17,767$
Delete any one digit $(1 - 9)$ of this difference, *but not 0*, and keep it secret.	17,6~~7~~6
Rearrange the digits of this number and tell me the resulting number	7, 6, 1, 6
I can tell you which digit you removed. It was a 7.	Mystified!

Figure 1.10

The important fact to know here is that the difference of two numbers with the same digits will always have a digit sum that is a multiple of 9 – that is, 9, 18, 27, 36, 45, and so on. So, when the trickster is told which digits remain after the player deletes one digit, the trickster simply takes the sum of the digits and selects the digit that would bring that digit sum to the next multiple of 9. That inevitably would be the digit that was removed. So, you now see that this "trick" depends completely on the notion that the difference of two numbers with the same digits must have a digit sum that is a multiple of 9. Such arithmetic trivia gives one a better grasp of number properties.

GUESSING-YOUR-NUMBER TRICK NO. 1

You may not even need a piece of paper for this trick. Ask your friend to select any number (a small number may be easier for quicker calculation) and have your friend add 1 to the number and then double that sum. To the resulting number have your friend add his secretly selected number +1 and then tell you the results. Whereupon you will now be able to very quickly calculate the original number your friend selected. All you need to do is subtract 3 and then divide that result by 3.

Just for practice, let's try one example using the secret number 18. First, we tell our friend to add 1 and double that number: $18 + 1 = 19 \times 2 = 38$. Next, we ask your friend to add the original secret number and then add 1 to that sum: $38 + 18 + 1 = 57$. When your friend tells his or her result, 57, we then subtract 3 and divide by 3, so that $57 - 3 = 54$ and then divide this result by 3 to get the original number, 18. Your friend will be astonished that you got the right number.

GUESSING-YOUR-NUMBER TRICK NO. 2

This is a simple and quick number trick to impress your friends at dinner while waiting for the food to be served. Have your friend secretly select any three-digit number. Then have your friend reverse the digits and subtract the two numbers. Now ask your friend to tell you the units digit of the final result. You will then be able to tell him or her the entire resulting number.

For example, suppose your friend selects 673. Reversing the digits and subtracting, we get $673 - 376 = 297$. When your friend tells you the units digit is 7, you should be able to say that the hundreds digit is 2 and the tens digit is 9. You do this by knowing that the units digit and the hundreds digit must add up to 9 and the tens digit will always be a 9. Therefore, since our exposed units digit was 7, the hundreds digit had to be 2 and the tens digit will always be a 9. Hence, we ended up with 297. Naturally, you could have asked your friend to expose the units digit, and that would allow you to get the number as well, since similarly the sum of the units and hundreds digits must always be 9.

GUESSING-YOUR-NUMBER TRICK NO. 3

Once again, using small numbers, your friend may not need a piece of paper for this trick. Begin by asking your friend to select the number, double the number, then add 5, and then multiply this result by 5. When your friend tells you the results of his calculation, all you need to do is delete the units digit and then subtract 2, and you will have the number that your friend originally selected.

When we try this trick with 36, our secret number, we double it to get 72, add 5 to get 77, and multiply this by 5 to get 385. This is the number we are given by our friend. From it we then delete the units digit to get 38, and when we subtract 2, we get our secret number 36.

GUESSING-YOUR-NUMBER TRICK NO. 4

Begin this trick by having your friend secretly select any number and multiply it by 4. Your friend then takes half of the result and multiplies it by 7.

Your friend then divulges his results, whereupon you divide that by 14 to get your friend's secret number.

To make this clearer, we can try this trick with the secret number 28. We multiply it by 4 to get 112 and then take half of that to get 56. We then multiply this by 7 to get 392. Finally, we divide this number by 14, which allows us to get the initial number, 28.

GUESSING-YOUR-NUMBER TRICK NO. 5

Once again, we ask our friend to select a secret number and double it. The friend is then asked to select any even number, which is to be added to this result. You instruct your friend to take half of this new number and multiply that by 4. The friend is then asked to subtract double their even number from this result and divulge this new result to you. To guess the original secret number, all you need to do is divide this new number by 4 and you will have the original secret number.

When we practice this trick with 23, we double it to get 46 and then select a random even number, say 18, which we add to the 46 to get 64. We then multiply 64×4 to get 256. Then we take one-half of 256 to get 128 and subtract from that twice our earlier selected even number 18, so that we have $128 - 36 = 92$. This number is then told to the trickster, who simply divides it by 4 to get the original secret number 23.

GUESSING-YOUR-NUMBER TRICK NO. 6

This trick will require somewhat more extensive calculation, since you will ask your friend to multiply his or her secret number by 50 and then add 72 to that number. From this result your friend is to subtract 111 and then add 39. This result is then to be divided by 5. Once this last number is given to you, you will be ready to do a simple calculation: take one-tenth of that number. You then will have exposed your friend's secret number.

We can try this trick with 7, which we then multiply by 50 to get 350 and then add 72 to get 422. From this result we subtract 111 to get 311 and then add 39 to get 350, which we then divide by 5 to get 70. As a trickster, all we need do is to take one-tenth of that number, which gives us the secret number, 7.

GUESSING-YOUR-NUMBER TRICK NO. 7

This unusual trick requires your friend to do various steps with her secret number. She will determine the steps while moving along, and tell you what calculation she does each time, yet without divulging the secret number or the result of each of these calculations.

Perhaps the best way to explain this trick is to use an example. Suppose your friend selects 8 as her secret number. She is then told to multiply the secret number by any number she chooses, but to tell you only what that multiplier is, but not the result. Once again, she can multiply that result by any other randomly selected number, which she divulges, but again not the result of the calculation. She then can divide the result by any number she chooses, telling you the divisor she chose, but not the resulting quotient. Now she can choose either to multiply or divide the result by another number, which she does and again tells you what that number is, but not the result of that calculation. She can do this any number of times using multiplication and division in any combination, yet each time telling you just the multiplier or divisor, not the result of the calculation. At a certain point she stops, and that is the number that she will divulge to you, the trickster.

All you have to do to get the original secret number is to do each of her calculations (with her divisors and multipliers), which you should have listed in order, but begin your calculation with 1. When you divide your friend's result by yours, you will have the secret number.

A simulation of this would probably be helpful to get a good grasp of the trick. Remember, we can use any combination of multiplications and divisions and divulge only the number we select as divisor or multiplier. We can begin with the secret number 3 and then *multiply* it by 7 to get 21. We tell the trickster that our first calculation was multiplying by 7. Next, we *multiply* our result, 21, by 5 to get 105, and again tell the trickster we just did a multiplication by 5. Now we will choose to *divide* our result by 2 to get 52.5. We then choose to do another *multiplication*, this time by 5, telling the trickster that we are multiplying by 5 to get 262.5. We can continue this combination of multiplications and divisions as long as we wish, but each time divulging the operator and the operation. At the end we need to be told what the end result is, in this case 262.5.

To get the original number, we need to do all the same calculations, but beginning with 1. The first calculation that the friend did was to *multiply* the secret number by 7, which in our case would yield 7. The next calculation that the friend did was to *multiply* by 5, which gives 35. The friend then *divided* by 2, and when we do it, we get 17.5. The next calculation was to *multiply* by 5 again to get 87.5. The friend chose to stop at this point and told us her result was 262.5. Now all we need to do to get the secret number is divide 262.5 by 87.5, and we get 3. How's that for a surprising trick!

GUESSING-YOUR-NUMBER TRICK NO. 8

This trick works in reverse from those previously presented. Here you will have your friend guess a number that you have selected in secret. You begin the trick by writing any number from 1 to 99 on a little slip of paper and folding it so that your friend cannot see it. Let him put it in his pocket. Then tell your friend to write down any number between 50 and 100, without letting you see it. Since you recall the number you wrote on the sheet of paper that is now in his pocket, you subtract that number from 99 and have your friend add that number to the number he secretly selected. Have your friend then delete the hundreds digit of the number thus obtained and add it to this number. Then tell your friend to subtract this newly obtained number from his original selected number. He is now ready to look into his pocket at the number you wrote on a little slip of paper. Much to his amazement, the number he has reached through this strange calculation is precisely the one you first wrote on that little slip of paper.

To better understand this trick, we will simulate it now with the following example. Suppose you write **31** on a little slip of paper, fold it up, and give it to your friend to put in his pocket without looking at it. Then you tell your friend to select any number between 50 and 100 without divulging it to you. Let's assume he chooses 67. In the meantime, you will subtract your hidden number of 31 from 99 to get 68, which you will have your friend add to *his* secret number. He then adds: $68 + 67 = 135$. He is then asked to delete the hundreds digit, 1, and add it to the remaining number: $35 + 1 = 36$. Now tell him to subtract this number from his original hidden number: $67 - 36 = 31$. How surprised he will be when he finds that the number he calculated is essentially the same one you wrote on a little slip of paper and which he hasn't seen until now.

This trick can be expanded to larger numbers as well. For example, you could ask your friend to write any number between 200 and 1000. You would then need to write a number between 100 and 200 on a little slip of paper, and instead of subtracting this number from 99, as we did in our previous example, you would subtract it from 999.

GUESSING-YOUR-NUMBER TRICK NO. 9

Here is a simple number trick that requires work by the trickster and the friend on whom the trick is being played. As trickster, have your friend secretly select any two-digit number, add 3 to that number, and then subtract 3 from the original number, writing both numbers on a piece of paper so that you cannot see either number. The next step is to square both numbers and then subtract the two squares. Your friend then tells you the result. Now you need to do some calculation to tell him what the original number was.

As trickster, you now need to do the following: double 6, which was the number you asked your friend to add. Then subtract half of 6, or 3, from his selected number, and divide the number your friend has given you by 12 to get the secret number.

It is always helpful to see the trick being played out to fully appreciate its effect. To do that, you, the trickster, ask your friend to select a secret number. Suppose she secretly selects 52. You now tell your friend to subtract 3 from her number to get 49, and to add 3 to it to get 55. By squaring these two numbers she will get 2401 and 3025, respectively. Taking the difference of these two numbers: $3025 - 2401 = 624$. This number your friend now divulges to you. Now, as trickster, you need to do some calculation to discover your friend's secret number. Divide 624, which your friend gave you, by 12 to get 52, which is the secret sought-after number.

GUESSING-YOUR-NUMBER TRICK NO. 10

The trickster asks the player to select any four-digit number. The player then has to add the digits and write that sum on the side. The trickster then asks the player to delete one digit from the original four-digit number. The trickster at the end will determine which number was deleted. From this newly formed three-digit number the player should subtract the digit sum from the original four-digit number. This number is then revealed to the trickster, who quickly takes the sum of these digits and subtracts it from the next higher multiple of 9. The result is the digit that was deleted.

Here is an example of how the trick would work with 1738 selected by the player, who then writes down 19, which is the sum of the digits of his original number. The player eliminates 7 from the original number. This leaves the three-digit number 138, from which the earlier digit sum (19) is to be subtracted to get 119. The trickster then takes the sum of these digits, 11, and subtracts it from the next higher multiple of 9, which is 18, to get 7. This number was the digit the player originally removed. This should surely amaze the player!

GUESSING-YOUR-NUMBER TRICK NO. 11

Which square number is the product of four consecutive odd integers? Done algebraically, the equation would evolve as follows:

$(n)(n + 2)(n + 4)(n + 6) = k^2$. This could then be converted to the following equation: $\left(n^2 + 6n + 4\right)^2 = k^2 + 16$. We know that k is odd, since it is the product of two consecutive odd integers; therefore, k^2 also must be odd. The only two numbers that can satisfy the right side of the equation as a square are 0 and 9. But since we need an odd number, the answer we seek is 9. And thus, we have $9 = (-3)(-1)(1)(3)$, which is what we sought.

GUESSING-YOUR-NUMBER TRICK NO. 12

The trickster should pose the following question to the audience. Multiply 93×107 mentally. After pausing a moment, the trickster should show how this can be done by representing the two numbers as $(100 - 7)(100 + 7) = 100^2 - 7^2 = 10,000 - 49 = 9951$, which is easily calculated mentally. This can be extended to other similar situations where we find that the relationship $(x - y)(x + y) = x^2 - y^2$ applies.

GUESSING-YOUR-NUMBER TRICK NO. 13

Pose the following challenge to your audience: Think of a number from which you subtract 7, and then multiply this difference by 7. You then amaze them by telling them that the resulting number should be the same as the original number reduced by 11 and then multiplied by 11.

The results of the two subtractions and multiplications should yield a number with factors 7 and 11. The number we seek, therefore, is 18. We can check this as follows: $18 - 7 = 11$, and $11 \times 7 = 77$. Now with 11, we get $18 - 11 = 7$, and $7 \times 11 = 77$. In general, the solution of this procedure $(x - n)(n) = (x - m)(m)$ is $(x = n + m)$, which is how we obtained 18 as $7 + 11$.

GUESSING-YOUR-NUMBER NO. 14

A trickster can offer the following challenge to the audience. They are to find the smallest number, which, when divided by each of the integers 2, 3, 4, 5, 6, 7, 8, 9, and 10, will yield in each case a remainder that is 1 less than the divisor.

The trickster needs to find the least common multiple of the numbers. This can be done by taking the factors of each number and then ensuring they are represented in the least common multiple, which here is $2^3 \times 3^2 \times 5 \times 7 = 2520$. Were we to divide this number by 2, 3, 4, 5, 6, 7, 8, 9, and 10, the remainder would be 0. To get a remainder of 1 less than each of the divisors, we simply need $2520 - 1 = 2519$. As an example, we can test this with any of the divisors, such as $2520 \div 7 = 359$ remainder of 6. Besides being a neat trick, this experience will give the audience some precious insight into number relationships.

GUESSING-YOUR-NUMBER TRICK NO. 15

We offer here a trick to see if your audience has some insight into simple arithmetic. In the following addition problem, each of the letters represents a different digit. Your audience's task is to find out which numerals are represented by the various letters.

$$
\begin{array}{r}
X\,Y \\
+\,Y\,Z \\
\hline
X\,X\,X
\end{array}
$$

First of all, the trickster and the audience should realize that the largest three-digit number that can be generated by adding two two-digit numbers is 198. Therefore, it should be clear that $X = 1$. Since the sum of these two-digit numbers is 111, and the first of the two numbers has 1 as a tens digit, it's easy to figure out that $Y = 9$ and $Z = 2$. So, $19 + 92 = 111$.

THE TRICK OF GUESSING RIGHT OR LEFT

This trick will give you some insights into the parity of numbers. When you play this game with a friend, give her a nickel and a dime and have her put the nickel in one hand and the dime in the other. Then ask your friend to multiply the value of the right-hand coin by 4, 6, or 8 and that of the left-hand coin by 3, 5, or 7. Ask your friend to add the two values arrived at and tell you what the result is. You will now be able to tell which coin was in her right hand and which in her left hand. If the number your friend tells you is *even*, then the nickel was in her right hand. If the number she tells you is *odd*, then the nickel was in her left hand. This can also be a good exercise for working with the parity of numbers—even and odd.

GUESS-THE-DIGIT TRICK

Here is a cute little trick with simple numbers as well as easy calculation. The trickster asks the player to select a number from 1 to 10, then multiply that number by the next larger number. From this product the player subtracts the smaller of the two original numbers. This result is multiplied by the smaller of the two original numbers. Now ask the player to tell you what the units digit of the final result is. Using the chart in Figure 1.11, locate the units digit of this resulting number in the upper row. The number below that will be the smaller of the two originally selected numbers.

Units digit	1	2	3	4	5	6	7	8	9	0
Smallest original number	1	8	7	4	5	6	3	2	9	10

Figure 1.11

As an example, the player selects 7 and then multiplies it by the next larger number, 8, to get $7 \times 8 = 56$. The player then subtracts the smaller of the two original numbers from 56 to get $56 - 7 = 49$. The player is then instructed to multiply this number, 49, by the smaller of the two original numbers, 7, to get $49 \times 7 = 343$. The player then tells what the units digit is (3). The trickster then looks at the chart and finds that the 7 in the lower row is below the 3 in the top row; therefore, the smaller of the two numbers originally selected is 7.

GUESSING YOUR NUMBER SEQUENCE

This number trick should surely impress the audience with your talent in guessing a number sequence. It involves asking your audience to select any three *consecutive* digits less than 50. Now, ask your audience to add the numbers together. They should add any multiple of 3 to the sum and inform you of which multiple of 3 they used. Next, have them multiply the number obtained by 67. Have them tell you the last two digits of this just-obtained number. You then will be able to determine their original sequence.

To find the sequence, you, as trickster, take the previously selected multiple of 3 that was given to you and divide it by 3. Then you add 1 to this quotient and subtract the result from the number determined by the last two digits of the player's final result. The result of this subtraction will yield the first of the three sequential numbers originally selected by the player.

Let us explain this trick through an illustrative example. Suppose your audience secretly selects the sequence 7, 8, 9. Your task as trickster is to discover the sequence. First, the player adds the numbers to get the sum of 24, to which they then add a multiple of 3, say 12, to get 36. Next, they are told to multiply this number by 67 to get $36 \times 67 = 2412$. We are then informed that the last two digits are 12. All we need to do now is divide the previously selected multiple of 3, which was 12, by 3, giving us 4, to which we then add 1 to get 5. Now we just subtract 5 from the number determined by the last two digits, $12 - 5 = 7$, which is the first number of the sequence that your audience originally selected.

BAFFLING YOUR AUDIENCE WITH A SURPRISE

When dividing 59 by 10 it is clear that we get 5 with a remainder of 9. This trick asks your audience to find a number that when divided by 10 leaves a remainder of 9, and when divided by 9 leaves a remainder of 8. This number will also leave a remainder of 7 when divided by 8. This pattern continues until when this number is divided by 3, it will leave a remainder of 2, and finally, when divided by 2, it will leave a remainder of 1. The challenge for your audience is to find a number that has all these characteristics.

One such number is 14,622,042,959. However, it is unrealistic to expect the audience to come up with this number. Therefore, there must be a smaller number you could have your audience search for, such as 3,628,799, which also satisfies the original challenge. Yet, the smallest solution can be obtained by looking for the least common multiple of 1, 2, 3, 4, …,8, 9, 10, which is $2^3 \times 3^2 \times 5 \times 7 = 2520$ and then subtract 1 to get 2519, the smallest number that satisfies the criteria.

MAXIMIZING A PRODUCT

Here's a simple problem that a trickster can give to see if his audience can think logically. They are challenged to find the largest product formed by two five-digit numbers, where no digit is used more than once across the two numbers. Logical thinking will prevail if the audience realizes that the largest numerals must be at the left end of each number. The desired product will be: $86,420 \times 97,531 = 8,428,629,020$. Notice that the two numbers in the product use all 10 digits exactly once. You should also know a pattern that shows how these two numbers were developed to create the largest product. (Hint: descending alternating digits crossed the two members of the product.)

THE TRICK OF GUESSING BROTHERS AND SISTERS

Suppose a friend comes from a large family of several brothers and sisters. You can tell her that you have a technique for determining how many of each she has. Tell her to follow these instructions:

1. Write down the number of brothers she has, but don't divulge it.
   We will assume she has 4 brothers.
2. Ask her now to double this number and add 1.
   We would get $2 \times 4 + 1 = 9$.
3. This result should be multiplied by 5.
   We then have $9 \times 5 = 45$.
4. Now ask her to add the number of sisters she has.
   We will assume here that she has 2 sisters. Therefore, in her calculation she will now have $45 + 2 = 47$.
5. When she tells you she got 47, you simply subtract 5.
   Therefore, we now have $47 - 5 = 42$.
6. From this two-digit number, you can tell her the tens digit is the number of brothers and the units digit the number of sisters.
   So, for our example, from **42** we deduce that she must have **4** brothers and **2** sisters.

A TRICK TO DETERMINE THE NUMBER OF FRIENDS AND ACQUAINTANCES

Here the player is asked how many friends and how many acquaintances he has, and the total number of former friends and former acquaintances he has had, but does not divulge any of this information. The player writes the number of friends and acquaintances on a hidden piece of paper. The trickster will now determine these numbers with the following calculations:

1. The player multiplies the number of current friends by 2.
2. The player then adds 3 to this number and multiplies it by 5.

3. Now the player adds this result to the number of current acquaintances and multiplies it by 10.
4. The player then adds this last number to the number of former friends and former acquaintances he has had and subtracts 150 from that number.
5. When the player divulges his final number, the trickster will use it to determine the number of current friends and acquaintances and the number of former friends and acquaintances.
6. This information is obtained as follows: the hundreds digit gives the number of current friends, the tens digit the number of current acquaintances, and the units digit the number of former friends and acquaintances.

Let's try this trick with the following example. Suppose the player indicates that he has 7 friends and 4 acquaintances. The player also indicates that he or she has 8 former friends and acquaintances. Using the five-step procedure above, we must do the following:

1. The player is to multiply the number of friends by 2: $7 \times 2 = 14$.
2. Then he adds 3 and multiplies this number by 5: $(14 + 3) \times 5 = 85$.
3. Next, the player adds the number of acquaintances to this last number $(4 + 85 = 89)$ and multiplies by 10 to get 890.
4. The player adds to this number (890) the number of former friends and acquaintances (8) and subtracts 150 to get $890 + 8 - 150 = \textbf{748}$, which he then tells the trickster.
5. From this number, 748, the trickster now knows that there are **7** friends, **4** acquaintances, and **8** former friends and acquaintances.

This trick will surely amaze the player, and perhaps prompt an investigation into the mathematics behind it.

A TRICK TO GUESS YOUR FRIEND'S AGE

For this trick we have a small stipulation: we need to assume that your friend's age is more than 10. In that case, have your friend add 90 to his or her age and then remove the hundreds digit and add it to the remaining number. When your friend tells you the result, you merely add 9 to get his or her age.

Let's try this with your friend who is 48 years old, but does not divulge it to you. First, she adds 90 to get 138. Removing the hundreds digit (1) and adding it to the remaining number (38) we get 39. When you are told that the remaining number is 39, you merely add 9 to get 48, which is the correct age!

HOW A TELEPHONE NUMBER CAN REVEAL A BIRTH YEAR

Guessing someone's birth year using this trick involves a multistep process. Begin by asking your friend, whose birth year you will try to discover, to write the last four digits of his telephone number without telling you. Then have him scramble these digits and subtract these two numbers. If the result has more than a one-digit number, he will add the digits to get a new number. If this result is still more than a one-digit number, he will add the digits once again, and continue this process until he reaches a single-digit number. Once a single-digit number has been reached, 8 is to be added to this digit. Then he is to add this number to the last two digits of his birth year. Subtracting 17 from this result will produce the birth year. Therefore, your friend's telephone number has revealed their birth year as well as his or her age!

Let's see how this trick works on the secret birth year of 1948. Remember, we're trying to see how your friend's telephone number can reveal the year 1948. Let's apply this trick to the telephone number $212 - 567 - 7879$. By scrambling the last four digits we get 9877. Subtracting these two numbers: $9877 - 7879 = 1998$. We now add the digits of this difference: $1 + 9 + 9 + 8 = 27$. Since we still don't have a single-digit number, we will once again add the digits of this sum: $2 + 7 = 9$. To this number we will add 8, to get $9 + 8 = 17$. We will now add the last two digits of our birth year: $48 + 17 = 65$. When we subtract 17 from this number 65, we will get 48; so, the birth year is 1948.

This trick is rather easy to explain. It works as it does because, when we take any number, rearrange the digits, and then subtract the two resulting numbers, the result will always be a number that is divisible by 9. Earlier, we found that when a number is divisible by 9, the sum of its digits has to be divisible by 9 as well. Therefore, we knew that adding 17 to the birth year and then subtracting that same amount was going to get us back to the birth year.

HOW TO DETERMINE YOUR FRIEND'S BIRTHDAY

Without your knowing the month and day on which your friend was born, ask him to do the following and he will end up with his birthday. Begin by having your friend multiply the day by 3, then add 5 to that product, and then multiply that sum by 4. Next, add to this result the day plus the month of the birthday and then subtract 20 from that result. Have your friend now tell you this result, which you will divide by 13. The resulting quotient will be the day, and the remainder will be the month.

Let's try this with an actual birthday, October 18.

Following the rules of the trick indicated above, we perform the following calculations:

1. Multiply the day by 3 to get $18 \times 3 = 54$.
2. Add 5 to that product: $54 + 5 = 59$.
3. Then multiply by 4: $59 \times 4 = 236$.
4. We then add to this number the month and day being sought: $236 + 18 + 10 = 264$.
5. Then we subtract 20 from that number: $264 - 20 = 244$.
6. When this number is divulged, we divide it by 13: $244 \div 13 = \mathbf{18}$ with remainder $\mathbf{10}$. This indicates that the birthday is the eighteenth day of the tenth month, or October 18.

A TRICK TO DETERMINE YOUR FRIEND'S AGE AT NEXT BIRTHDAY

This trick involves several repeated calculations, albeit simple ones, to determine your friend's birthdate and her age on that day. You will ask your friend to do the following calculations. When you get the end result, you will do one simple subtraction and have the required information: birthdate and age on the next birthday.

Here are the instructions to give to your friend.

1. Multiply the number of the month by 100.
2. To that number add 17 and double the result.
3. Add 5 to the last number and multiply by 10.
4. Add 23 to the last number, multiply it by 5, and add 49 to the result.
5. At this point you ask your friend to tell you the final number, and you subtract 365. You now obtain the age at the next birthday by taking the last two digits on the right as the age, the next two digits as the date, and the last one or two digits as the month.

Let's try this trick on someone who will be 49 years old next August 17.

1. Multiply the eighth month (August) by 100 to get 800.
2. Next, we add 17 and double the result: $800 + 17 = 817; 817 \times 2 = 1634$.
3. We now add 5 to that number and multiply it by 10: $1634 + 5 = 1639; 1639 \times 10 = 16,390$.
4. We add 23 to that number, then multiply it by 5 and add $4,916,390 + 23 = 16,413; 16,413 \times 5 = 82,065; 82,065 + 49 = 82,114$.

5. The result, 82,114, has been divulged to you. Now you subtract 365 from that number, $82,114 - 365 = 81,749$, which gives you the number providing you with the necessary information.

We then partition our final result (81,749) as 8/17 and 49, which indicates August 17, age 49. Remember, steps 1 through 4 are done by the friend, and step 5 shows your cleverness by subtracting 365 from your friend's final result to get the required information.

THE TRICK OF GUESSING YOUR FRIEND'S EXACT DATE OF BIRTH

Now we show a trick to get the exact birth date. Have your friend follow these instructions:

1. Begin, by asking him to write the number of the month in which he was born.
2. Now, have him add the next number in sequence to this number and multiply it by 5.
3. Next, have him tack on a 0 to the right of the number reached.
4. Ask your friend to select any number less than 100 and tell you what it is. Then he is to add that number to the previous one he calculated.
5. Now, ask your friend to add to this new number the day of the month on which he was born.
6. Once again, your friend is to select any number less than 100, tell you what it is, and tack it onto the right of the number he has arrived at so far.
7. To this result he is to add the number determined by the last two digits of his birth year.

When he tells you the number he has reached, you will be able to tell him the date of his birth. To do that you need to add 50 to the first number your friend brought into the picture (in step 4 above). Then place the number he told you in step 6 to the right of the number you just reached. Then take your friend's last number, which he told you, and subtract the number you just developed from it. That will give you your friend's birthdate.

To best understand this process, let's work an illustrative example. Without divulging the information to the trickster, we will use the birth date of October 18, 1942.

We begin by writing down the number of our secret birth month, 10. According to the instructions, we will add the next number in sequence and multiply by 5: $(10 + 11) \times 5 = 105$. Next, we are asked to attach a 0 to this number to get 1050. We now tell the trickster that we are adding 36 to our secret number to get 1086. Since our day of birth is the eighteenth, we add that to our previously obtained number to get $1086 + 18 = 1104$. Once again, we are asked to select a number less than 100, which we do with 28, and tell the trickster what that number is. Then we tack that number onto the right side of the number we previously reached, 1104, so that we get 110,428. We now add the last two digits of the birth year to this previously reached number: $110,428 + 42 = 110,470$.

Now the trickster begins the calculation. He needs to add 50 to the number that we first added into the process: $50 + 36 = 86$. Then he tags on the second number we selected, 28, onto this number to get 8628. He then subtracts this number from the last number to get $110,470 - 8628 = 101,842$, or in date form 10/18/42, which is October 18, 1942.

Although the process is a bit tedious, it is nonetheless an impressive trick. It will motivate further investigation to see why it works.

TRICKING YOUR FRIENDS TO EXPOSE A FIXED NUMBER

This is an easy trick that always ends up with a fixed number. The example will end up with 6. Follow the few steps needed to get to this magic number. Have your friend pick any number, add it to the next higher number, and then add 11 to that sum. When your friend divides the result by 2 and subtracts this number from the original one he selected, 6 should appear.

Let's try that by starting off with 618. When we add the next higher number to it, we get $618 + 619 = 1237$. We then add 11 to the sum: $11 + 1237 = 1248$. When we divide by 2, we have 624. When we subtract the original number (618), we get $624 - 618 = 6$, as predicted!

You might want to vary this trick so that you don't always end up at the same number, as we did here with 6. You can vary the answer by changing the number that you ask your friend to add to the sum of the first two numbers. In our case above, that number was 11. For example, if you had used 17 instead of 11 as the add-on number, your result would be 9.

Figure 1.12 provides a chart that tells you what results you will have when you vary the "add-on number." As you will notice, it is always an odd number.

The add-on number	3	5	7	9	11	13	15	17	19	21	23	25
The resulting number	2	3	4	5	6	7	8	9	10	11	12	13

Figure 1.12

EXPOSING THE FOUR KNOWN DIGITS OF YOUR SOCIAL SECURITY NUMBER

It currently is common practice to identify yourself online by providing the last four digits of your Social Security number. Here is a trick that you can use to expose the last four digits of your friend's Social Security number. To do this, have your friend secretly write down the last four digits of his Social Security number. Then ask him to write the first two digits as a two-digit number and add that number to the next higher number in sequence. Then ask him to multiply that sum by 5 and place a 0 to the right of the result, thereby creating a four-digit number with a units digit of 0. Ask your friend to select any number between 11 and 99 inclusive—telling you what it is—and add that number to the previously obtained four-digit number. The last step is to have your friend add the last two digits of the original Social Security number (as a two-digit number) to the previously obtained number and tell you what that number is.

Now, to determine his original number, add 50 to the number between 11 and 99 he reported to you, which he selected. Add it to the four-digit number he ended up with, and you will have discovered the last four digits of his Social Security number.

Once again, an example of this trick will help solidify understanding. Suppose your friend's Social Security number ends in 8632. He takes the first two digits as a number, 86, and adds it to its successor number 87 to get 173, then multiplies that by 5 to get 865. He then places a 0 on the right side of 865 to get 8650. Now, your friend is to select a number between 11 and 99. He selects 18 and tells you this number, which he then adds to the 8650 to get 8668. Then you ask him to add the last two digits of the original number to this number to get $8668 + 32 = 8700$. This number is then divulged to you, and you then subtract $50 + 18 = 68$ from the number to get 8632. These are the last four digits of his Social Security number.

GUESSING A CHILD'S NUMBER

This little trick might be easier to do with a calculator. Have a child choose a two-digit number, which he will not divulge to you. First you will ask the child to square this number. Then the child should take this original two-digit number, add 8 to it, and square that number as well. Then the child is to subtract the two squares that were just obtained. Now you have to do some work. Take 8, which you used earlier, and double it to get 16. Ask the child for the number he or she obtained at the end and divide it by 16. Then subtract half of the fixed number 8 from this result, and you will have arrived at the number the child originally selected.

Perhaps an example might clarify this procedure. Suppose the child secretly selects 48. By adding 8, he or she will arrive at 56. Squaring each of these numbers yields $28^2 = 2304$ and $56^2 = 3136$. Subtracting these two numbers yields $3136 - 2304 = 832$. Dividing this number by 16 (which is twice the 8 used earlier) yields 52. To get the number that the child originally selected, you merely subtract half of 8 from 52 to get 48, which is precisely the original number selected.

N.B. This trick does not depend on 8, and it can work nicely with any even number replacing the 8.

FINDING THE DELETED DIGIT

The cleverness that you can demonstrate with this little trick will surely impress your friends. Here you will discover a digit that your friend will delete from a four-digit number that she has selected. The trick works as follows. Ask your friend to select a four-digit number without telling you what it is. Then have your friend find the sum of the digits and retain this number. At this point ask your friend to delete one of the digits of the four-digit number she selected. The remaining three digits will determine a new three-digit number from which your friend should subtract the previously obtained sum of the digits of the original four-digit number. At this point your friend will tell you the number she obtained. To determine which digit was deleted, find the sum of the digits of the number just provided to you and subtract that number from the next higher multiple of 9. Be aware that if the sum of the digits of the number your friend provided to you is a multiple of 9, you will think that the deleted digit was a 0, but also could have been a 9. At this point, your best response to your friend is "I believe that your deleted number was either 0 or 9." Even that response will impress your friend.

To better understand this trick, we will consider a sample case of it. Suppose your friend selects the four-digit number 3652. Next, your friend will find the sum of the digits, which in this case is 16. At this point, your friend is to delete one of the digits. Let's assume that she deletes the 6. The remaining number is now 352, from which she is to subtract the previous digit sum, which was 16, to get $352 - 16 = 336$. This is the number that she is to reveal to you. Now you need to find the sum of the digits of 336, which is 12, and then subtract this number from the next multiple of 9, which is 18. You will get 6, which is the originally deleted number. This should certainly get a curious reception.

A MULTIPLICATION TRICK

Here we present a rather unusual method to multiply two numbers. It is actually quite simple, yet somewhat cumbersome. Consider finding the product of 43 × 92. As we work on this trick, we begin by setting up a chart of two columns with the two members of this multiplication task in the first row. Figure 1.13 shows the 43 and 92 heading up the columns. One column is formed by doubling each number to get the next number, while the other column takes half the previous number and drops the remainder if there is one. In this example, our left-side column will be the doubling column, and the right-side column will be the halving column. Notice that halving an odd number such as 23 (the third number in the right-side column) gives us 11 with a remainder of 1 and we simply drop the 1. The rest of this halving process should now be clear.

Doubling	Halving
43	92
86	46
172	23
344	11
688	5
1376	2
2752	1

Figure 1.13

We then identify the odd numbers in the halving column, then get the sum of the partner numbers in the doubling column, which are highlighted in bold type. This sum gives you the originally required product of 43 × 92. In other words, with this method we get 43 × 92 = 172 + 344 + 688 + 2752 = 3956.

In the example shown in Figure 1.13, we chose to have the left-side column (the doubling column) and the right-side column (the halving column). We could also have done this method by halving the numbers in the left-side column and then doubling those in the right-side column. See Figure 1.14.

To complete the multiplication, we find the odd numbers in the halving column (in bold type), and then get the sum of their partner numbers in the second column (now the doubling column). This gives us 43 × 92 = 92 + 184 + 736 + 2944.

Although it would be rather silly to use this method to do multiplication—especially since calculators are so readily available. However, it should be fun to observe how this primitive arithmetic algorithm actually does work. Explorations of this kind, not only are instructive, but also can be entertaining.

Halving	Doubling
43	92
21	184
10	368
5	736
2	1472
1	2944

Figure 1.14

Here you see what was done in the above multiplication algorithm. We define an *algorithm* as a step-by-step problem-solving procedure, especially an established, recursive computational procedure for solving a problem in a finite number of steps.

$$\star 43 \times 92 = (21 \times 2 + 1)(92) \quad = 21 \times 184 + 92 = 3956$$
$$\star 21 \times 184 = (10 \times 2 + 1)(184) = 10 \times 368 + 184 = 3864$$
$$10 \times 368 = (5 \times 2 + 0)(368) \quad = 5 \times 736 \ + 0 = 3680$$
$$\star 5 \times 736 = (2 \times 2 + 1)(736) \quad = 2 \times 1472 + 736 = 3680$$
$$2 \times 1472 = (1 \times 2 + 0)(1472) = 1 \times 2944 \ + 0 = 2944$$
$$\star 1 \times 2944 = (0 \times 2 + 1)(2944) = 0 + \underline{2944} = 2944$$

Column Total is 3956

For those familiar with the binary system (i.e., base 2), this method can also be explained by the following representation.

$$(43)(92) = (1 \times 2^5 + 0 \times 2^4 + 1 \times 2^3 + 0 \times 2^2 + 1 \times 2^1 + 1 \times 2^0)(92)$$
$$= 2^0 \times 92 + 2^1 \times 92 + 2^3 \times 92 + 2^5 \times 92$$
$$= 92 + 184 + 736 + 2944$$
$$= 3956$$

Whether or not you fully understand the discussion of this method of multiplication, you should at least now have a deeper appreciation for the easier multiplication algorithm you learned in school, even though most people today use calculators. There are many other multiplication algorithms, some of which we will explore in the next few tricks. The one shown next is perhaps one of the strangest, and through its strangeness we can appreciate the powerful consistency of mathematics that allows us to conjure up such an algorithm.

A TRICKY METHOD TO MULTIPLY TWO NUMBERS

Let's delve right into some examples of this tricky method of multiplying two-digit numbers and see why it works. Take, for example, the multiplication 95×97. The following steps can be done mentally (with some practice, naturally!):

Step 1: $95 + 97 = 192$

Step 2: Delete the hundreds digit $= 92$

Step 3: Tag two zeros onto the number $= 9200$

Step 4: $(100 - 95) \cdot (100 - 97) = 5 \times 3 = 15$

Step 5: Add the last two numbers $= 9215$, which is the desired product!

Here is another example of this technique:

$93 \times 96 = ...$

$93 + 96 = 189$

$\cancel{1}89$ (Delete the hundreds digit.)

Tack on two zeros $= 8900$

Then add $(100 - 93) \times (100 - 96) = 7 \times 4 = 28$ to the previously obtained

number, to get 8928, which is what we sought: 93×96.

This technique also works when seeking the product of two numbers that are further apart:

89×73

$89 + 73 = 162$

$\cancel{1}62$ (Delete the hundreds digit.)

Tag on two zeros $= 6200$

Then add $(100 - 89) \times (100 - 73) = 11 \times 27 = 297$ to get 6497,

which again leads us to the product of 89×73.

For those who might be curious as to why this unusual technique works, we can use simple algebra that will justify it. We begin with the two two-digit numbers:

$$(100 - a) \text{ and } (100 - b) (\text{where } 0 < a, \text{ and } b < 100).$$

Step 1: $(100 - a) + (100 - b) = 200 - a - b$

Step 2: Delete the hundreds digit—which means subtracting 100 from the number:

$$(200 - a - b) - 100 = 100 - a - b$$

Step 3: Tack on two zeros, which means multiply by 100:

$$(100 - a - b) \times 100 = 10,000 - 100a - 100b$$

Step 4: $a \times b$

Step 5: Add the last two results from steps 3 and 4, and then do some factoring:

$$10,000 - 100a - 100b + ab$$
$$= 100(100 - a) - (100b - ab)$$
$$= 100(100 - a) - b(100 - a)$$
$$= (100 - a)(100 - b), \text{ which is what we set out to show.}$$

Now you just need to practice this trick to master it!

A QUICK MULTIPLICATION TRICK

Some people are very adept at rapid multiplication. Here is a procedure used to do quick multiplication with a pair of two-digit numbers that will impress your audience. It is done as follows:

- Begin by multiplying the units digits of the two numbers. If a two-digit number results, write the units digit and carry the tens digit to the next step.
- Next, multiply the units digit of one number by the tens digit of the other number. Then multiply the tens digit of the first number by the units digit of the second number and add the two products. Then add any number carried over from the previous step. Place the units digit of the result to the left of the previously obtained single digit and carry the tens digit to be added to the next step's calculation.
- Lastly, multiply the two tens digits and add the number carried over from the previous step. Place this result to the left of the two previously obtained numbers, and you will have the final result of the multiplication.

At first sight this appears to be a rather complicated procedure. But after we work through an example, the procedure will become rather simple and clear and relatively fast to do. As our example, let us use this procedure to multiply 59×38.

- First, we multiply the units digits of the two numbers to be multiplied: $9 \times 8 = 72$. This gives us the units digit of our ultimate answer, **2**, and we will carry the 7 to the next step.
- We now multiply the units and tens digits of the two numbers: $5 \times 8 = 40$, and $9 \times 3 = 27$. We add these two numbers, $40 + 27 = 67$, and then add the 7 carried from the previous step to get $67 + 7 = 74$. We place the **4** to the left of the previously obtained **2** and carry the 7 to the next step. So far, we have the tens and units digits of our sought-after product, **42**.
- We now multiply the two tens digits to get $3 \times 5 = 15$, and add the carried 7 to get $15 + 7 = 22$. We now place this number to the left of the two previously obtained final digits to get 2242, which is the product of 59×38.

A quick review of this procedure is simply to multiply the pair of units digits, the cross product of tens and units digits, and then the pair of tens digits—each time carrying the tens digit as appropriate. To successfully use this technique you will need to practice it with a number of examples so that your calculating speed will become impressive. An ambitious reader may want to extend this technique to multiplying two-digit numbers by three-digit numbers, and three-digit numbers by other three-digit numbers.

THE ADDITION TRICK OF SIX NUMBERS

This trick is fun and, as before, shows how we can analyze a seemingly baffling result through simple algebra. We begin by asking our trusty friend to select any three-digit number with no two like digits (omitting the zero). Then, we have him make five other numbers with these same digits.[2] Suppose our friend selected 473, then made a list of all numbers formed from these digits:

$$473$$
$$437$$
$$347$$
$$374$$
$$743$$
$$734$$
———

The trick is to baffle your friend by getting the sum of these numbers, 3108, faster than he can even write the numbers. How can this be done so fast? All we do is get the sum of the digits of the original number (here $4 + 7 + 3 = 14$) and then multiply 14 by 222 to get 3108, which is the required sum. Why 222? Let us inspect some of the many quirks of number properties using simple algebra:

Consider $\overline{abc} = 100a + 10b + c$, where a, b, c are numbers from the set $\{1, 2, 3, ..., 9\}$. The sum of the digits is $a + b + c$. We now represent all of the six numbers formed by these digits on our list as:

$$100a + 10b + c$$
$$+100a + 10c + b$$
$$+100b + 10a + c$$
$$+100b + 10c + a$$
$$+100c + 10a + b$$
$$+100c + 10b + a$$

When we add these numbers, we get:

$$100(2a + 2b + 2c) + 10(2a + 2b + 2c) + 1(2a + 2b + 2c)$$
$$= 200(a + b + c) + 20(a + b + c) + 2(a + b + c)$$
$$= 222(a + b + c), \text{ which is 222 times the sum of the digits.}$$

If you really want to be slick and dramatically impress your friend, you might write the various digit sums, shown in Figure 1.15, on a small piece of paper for easy reference. That would simplify things so that you avoid the actual multiplication by 222.

Original digit sums	6	7	8	9	10	11	12	13	14	15
Sum of the six numbers	1332	1554	1776	1998	2220	2442	2664	2886	3108	3330
Original digit sums	16	17	18	19	20	21	22	23	24	
Sum of the six numbers	3552	3774	3996	4218	4440	4662	4884	5106	5328	

Figure 1.15

LOCATING THE NUMBER TRICK

This trick locates a number in an organized arrangement. Suppose we write the natural numbers in a triangle, as shown in Figure 1.16.

$$1$$

$$2 \quad 3 \quad 4$$

$$5 \quad 6 \quad 7 \quad 8 \quad 9$$

$$10 \quad 11 \quad 12 \quad 13 \quad 14 \quad 15 \quad 16$$

$$17 \quad 18 \quad 19 \quad 20 \quad 21 \quad 22 \quad 23 \quad 24 \quad 25$$

$$26 \quad 27 \quad 28 \quad \ldots$$

Figure 1.16

The trickster begins questioning his audience as follows: In which row would we find 2000? In which column would 2000 be located? We should first ask: Do we have enough information to answer the question? Will we actually have to enlarge the triangular arrangement of numbers to the point where 2000 appears? Here is a situation that will circumvent actual counting by looking for the number patterns.

Notice that each row ends in a square number whose square root corresponds to the number of the row; that is, the third row, for example, ends in 9 or 3^2. The fifth row ends in 25, which is 5^2; and so, the nth row would end in n^2.

The first number in each row is one greater than the last number of the previous row, which is the square of the number of that row. Therefore, for the row following the nth row, we can write the first number as $(n-1)^2 + 1$.

The middle number of each row is the average of the first and the last numbers of that row, since the numbers are consecutive. Hence, for the 9th row the middle term is in the nth column and is:

$$\frac{[(n-1)^2 + 1] + n^2}{2} = \frac{n^2 - 2n + 1 + 1 + n^2}{2} = n^2 - n + 1$$

Now to our quest to locate 2000. Since $44^2 = 1936$, and $45^2 = 2025$, 2000 must be between 44^2 and 45^2. Therefore, it must be in the row that ends in 45^2, which means it must lie in row 45. Using our formula for the first term of a row, $(n-1)^2 + 1$, we get the first term of row 45 to be $(45-1)^2 + 1 = 1937$. The last term is $45^2 = 2025$, and the middle number is the average of these two, $\dfrac{1937 + 2025}{2} = 1981$, and is in the 45th column. Therefore, we can then easily determine that $2000 - 1981 = 19$, which places 2000 in the $19 + 4$ or the 64th column. Although this trick might require some preparation, it still can impress others as well as show how some logical thinking can be most useful.

THE TRICK OF RETURNING TO A NUMBER

Select any three-digit number and write it twice to form a six-digit number. For example, if you choose 357, then write the six-digit number 357,357. We now—perhaps using a calculator—divide this number by 7. Then we divide the resulting quotient by 11, and lastly divide that quotient by 13 as follows:

$$\frac{357,357}{7} = 51,051$$

$$\frac{51,051}{11} = 4641$$

$$\frac{4,641}{13} = 357$$

This is the first number you started with!

This works as it does because to form the original six-digit number, you actually multiplied the original three-digit number by 1001. That is, $357 \times 1001 = 357,357$. However, $1001 = 7 \times 11 \times 13$. Therefore, by dividing successively by 7, 11, and 13, we have undone the original multiplication by 1001, leaving the original number. The audience, unaware of this, will be quite surprised at the result.

We can look at how 1001 can also help us multiply other number combinations.

$$221 \times 77 = (17 \times 13) \times (11 \times 7) = 1001 \times 17 = (1000 \times 17) + (1 \times 17) = 17,017$$

$$264 \times 91 = (24 \times 11) \times (13 \times 7) = 1001 \times 24 = (1000 \times 24) + (1 \times 24) = 24,024$$

$$407 \times 273 = (37 \times 11) \times (3 \times 7 \times 13) = 1001 \times 111 = (1000 \times 111) + (1 \times 111) = 111,111.$$

Yes, arithmetic can expose some hidden numerical treasures.

DIGIT SUMS TRICK

When we speak of the sum of the digits of a number, we simply add the digits. For example, the sum of the digits of 251 is simply $2 + 5 + 1 = 8$. Let's take a big leap and determine the sum of the digits of all the numbers from

1 to 1,000,000. One way to answer this question is to begin by summing the digits of consecutive numbers starting with 1 (see Figure 1.17).

The number	The Digit Sum
1	1
2	2
3	3
⋮	⋮
35	$3 + 5 = 8$
36	$3 + 6 = 9$
37	$3 + 7 = 10$
38	$3 + 8 = 11$
etc.	

Figure 1.17

This does not seem to be a very efficient, or elegant, way of proceeding. Rather than simply trying to add the digits of all the numbers, we will use a cleverer arrangement of the numbers to obtain the sum without actually doing all this tedious addition.

Let's consider the list of the numbers from 0 to 999,999 in two directions: ascending order and descending order. (See Figure 1.18.)

Ascending order	Descending order	Sum of the digits of each pair of numbers
0	999,999	54
1	999,998	54
2	999,997	54
3	999,996	54
⋮	⋮	⋮
127	999,872	54
⋮	⋮	⋮
257,894	742,105	54
⋮	⋮	⋮
999,997	2	54
999,998	1	54
999,999	0	54

Figure 1.18

We will leave 1,000,000 for a bit later. Since every pair of numbers has a digit sum of 54, we have 54 × 1,000,000 and the two columns have the same numbers, the sum of the digits in one column is 27 × 1,000,000. We must now add the digit sum of the last number, 1,000,000, which is 1.

Therefore, the digit sum of all numbers from 1 to 1,000,000 is 1,000,000 × 27 + 1 = 27,000,001.

From this example, you can see that arithmetic is more than just doing the basic operations—addition, subtraction, multiplication, and division. It requires a bit of thinking as well, and as a result provides a neat trick to entertain your audience.

UNUSUAL NUMBERS

There are numbers that for inexplicable reasons produce some amazing results. Your audience will be surprised at the results you will produce with some of these unusual numbers.

We begin with 76,923 and multiply sequentially by 1, 10, 9, 12, 3, and 4. Look at the results of these six multiplications, and you will see the most amazing relationship among them.

$$76,923 \times 1 = \mathbf{0}76,923$$
$$76,923 \times 10 = \mathbf{7}69,230$$
$$76,923 \times 9 = \mathbf{6}92,307$$
$$76,923 \times 12 = \mathbf{9}23,076$$
$$76,923 \times 3 = \mathbf{2}30,769$$
$$76,923 \times 4 = \mathbf{3}07,692$$

Notice how the first digits of the numbers generated show the original number 076,923. Furthermore, notice how the numerals sequence through each of the products presented. The diagonal through these product numbers from upper right to lower left consists of all 3's. If you inspect the products further, you will see patterns formed by parallels to the diagonal just highlighted.

When multiplied by other numbers, such as 2, 7, 5, 11, 6, and 8, 76,923 provides another striking pattern.

$$76,923 \times 2 = \mathbf{1}53,846$$
$$76,923 \times 7 = \mathbf{5}38,461$$
$$76,923 \times 5 = \mathbf{3}84,615$$
$$76,923 \times 11 = \mathbf{8}46,153$$
$$76,923 \times 6 = \mathbf{4}61,538$$
$$76,923 \times 8 = \mathbf{6}15,384$$

Once again, we notice that the first digits of these products exhibit the same number as the first product, 153,846. Here as well the diagonal from the upper right to the left of the products is represented by 6, and parallel lines are also represented by the same digits.

Another number that yields some symmetric results is obtained when we divide 999,999 by 7, which is 142,857. This time we will multiply 142,857 sequentially by 1, 3, 2, 6, 4, and 5 as shown below:

$$142,857 \times 1 = 142,857$$
$$142,857 \times 3 = 428,571$$
$$142,857 \times 2 = 285,714$$
$$142,857 \times 6 = 857,142$$
$$142,857 \times 4 = 571,428$$
$$142,857 \times 5 = 714,285$$

Again, you see the first digits of each of the products to be 1, 4, 2, 8, 5, and 7, and the diagonal from upper right to lower left consists of all 7's. Patterns similar to those previously shown are embedded in this group of products. Such unusual number patterns will always be well received by your audience.

TRICKS EVOLVING FROM NUMBER PECULIARITIES

As a trickster you may want to show some of the following number peculiarities (Figures 1.19–1.26):

987654321	×	9	=	8 888 888 889
987654321	×	18	=	17 777 777 778
987654321	×	27	=	26 666 666 667
987654321	×	36	=	35 555 555 556
987654321	×	45	=	44 444 444 445
987654321	×	54	=	53 333 333 334
987654321	×	63	=	62 222 222 223
987654321	×	72	=	71 111 111 112
987654321	×	81	=	80 000 000 001

Figure 1.19

```
        0  ×  9  +  1  =  1
        1  ×  9  +  2  =  11
       12  ×  9  +  3  =  111
      123  ×  9  +  4  =  1,111
    1,234  ×  9  +  5  =  11,111
   12,345  ×  9  +  6  =  111,111
  123,456  ×  9  +  7  =  1,111,111
1,234,567  ×  9  +  8  =  11,111,111
12,345,678 ×  9  +  9  =  111,111,111
```

Figure 1.20

```
999,999  ×   1  =  0,999,999
999,999  ×   2  =  1,999,998
999,999  ×   3  =  2,999,997
999,999  ×   4  =  3,999,996
999,999  ×   5  =  4,999,995
999,999  ×   6  =  5,999,994
999,999  ×   7  =  6,999,993
999,999  ×   8  =  7,999,992
999,999  ×   9  =  8,999,991
999,999  ×  10  =  9,999,990
```

Figure 1.21

```
         11  ×           11  =                   121
        111  ×          111  =                 12,321
      1,111  ×        1,111  =              1,234,321
     11,111  ×       11,111  =            123,454,321
    111,111  ×      111,111  =         12,345,654,321
  1,111,111  ×    1,111,111  =      1,234,567,654,321
 11,111,111  ×   11,111,111  =    123,456,787,654,321
111,111,111  ×  111,111,111  =  12,345,678,987,654,321
```

Figure 1.22

```
        9  ×          9  =  81
       99  ×         99  =  9,801
      999  ×        999  =  998,001
    9,999  ×      9,999  =  99,980,001
   99,999  ×     99,999  =  9,999,800,001
  999,999  ×    999,999  =  999,998,000,001
9,999,999  ×  9,999,999  =  99,999,980,000,001
```

Figure 1.23

$$\begin{array}{rcrcrcl}
& \times & 9 & + & 8 & = & 8 \\
9 & \times & 9 & + & 7 & = & 88 \\
98 & \times & 9 & + & 6 & = & 888 \\
987 & \times & 9 & + & 5 & = & 8{,}888 \\
9{,}876 & \times & 9 & + & 4 & = & 88{,}888 \\
98{,}765 & \times & 9 & + & 3 & = & 888{,}888 \\
987{,}654 & \times & 9 & + & 2 & = & 8{,}888{,}888 \\
9{,}876{,}543 & \times & 9 & + & 1 & = & 88{,}888{,}888 \\
98{,}765{,}432 & \times & 9 & + & 0 & = & 888{,}888{,}888
\end{array}$$

Figure 1.24

$$37037 \times 3 = 111111$$
$$37037 \times 9 = 333333$$
$$37037 \times 6 = 222222$$
$$37037 \times 18 = 666666$$
$$37037 \times 12 = 444444$$
$$37037 \times 15 = 555555$$

Figure 1.25

$$\begin{array}{rclcl}
1 & = & 1 & = & 1^2 \\
1+2+1 & = & 2+2 & = & 2^2 \\
1+2+3+2+1 & = & 3+3+3 & = & 3^2 \\
1+2+3+4+3+2+1 & = & 4+4+4+4 & = & 4^2 \\
1+2+3+4+5+4+3+2+1 & = & 5+5+5+5+5 & = & 5^2 \\
1+2+3+4+5+6+5+4+3+2+1 & = & 6+6+6+6+6+6 & = & 6^2 \\
1+2+3+4+5+6+7+6+5+4+3+2+1 & = & 7+7+7+7+7+7+7 & = & 7^2 \\
1+2+3+4+5+6+7+8+7+6+5+4+3+2+1 & = & 8+8+8+8+8+8+8+8 & = & 8^2 \\
1+2+3+4+5+6+7+8+9+8+7+6+5+4+3+2+1 & = & 9+9+9+9+9+9+9+9+9 & = & 9^2
\end{array}$$

Figure 1.26

Here is one last pattern with which you could entertain your audience. While playing with 9, you might try to have your audience find an eight-digit number in which no digit is repeated, and which, when multiplied by 9, yields a nine-digit number in which no digit is repeated. Here are a few such possibilities:

$$
\begin{array}{rcccl}
81{,}274{,}365 & \times & 9 & = & 731{,}469{,}285 \\
72{,}645{,}831 & \times & 9 & = & 653{,}812{,}479 \\
58{,}132{,}764 & \times & 9 & = & 523{,}194{,}876 \\
76{,}125{,}483 & \times & 9 & = & 685{,}129{,}347 \\
\end{array}
$$

Figure 1.27

PICK YOUR FAVORITE NUMBER

In this trick you present your audience with the following number and ask them to circle their favorite digit in it: 12,345,679. You will then ask them to do a calculation producing a number repeating this numeral nine times. (Notice that the 8 is missing.) Suppose they encircle 6. All they need to do is multiply $6 \times 9 = 54$ by 12,345,679, and their favorite numeral will appear nine times as 666,666,666. Had they chosen 4, they would multiply $4 \times 9 = 36$, then multiply $36 \times 12{,}345{,}679$, and their favorite numeral would appear as 444,444,444. The chart in Figure 1.28 shows you how those multiplications allow them to generate their favorite number.

$$
\begin{array}{rcrcl}
12345679 & \times & 9 & = & 111{,}111{,}111 \\
12345679 & \times & 18 & = & 222{,}222{,}222 \\
12345679 & \times & 27 & = & 333{,}333{,}333 \\
12345679 & \times & 36 & = & 444{,}444{,}444 \\
12345679 & \times & 45 & = & 555{,}555{,}555 \\
12345679 & \times & 54 & = & 666{,}666{,}666 \\
12345679 & \times & 63 & = & 777{,}777{,}777 \\
12345679 & \times & 72 & = & 888{,}888{,}888 \\
12345679 & \times & 81 & = & 999{,}999{,}999 \\
\end{array}
$$

Figure 1.28

THE 6174 TRICK

A quirk in our decimal number system enables us to do this most amazing trick. Here you will have your audience select any four-digit number in which the digits are different, and you enable them all to end up with 6174. To streamline this process you could ask your audience to use a calculator, unless they are proficient in subtraction. Have them follow these steps.

1. *Select a four-digit number—except one that has all digits the same.*
2. *Rearrange the digits of the number so that they form the largest number possible. (In other words, write the number with the digits in descending order.)*
3. *Then rearrange the digits of the number so that they form the smallest number possible. (That is, write the number with the digits in ascending order. Zeros can take the first few places.)*

4. *Subtract these two numbers (the smaller from the larger).*
5. *Take this difference and continue the process, repeatedly. Don't stop until something unusual happens.*

Eventually your audience members will individually arrive at 6174—perhaps after one subtraction, or after several subtractions. Once they arrive at 6174, they will find themselves in an endless loop. This means that continuing the process with 6174 will keep producing 6174 as the result. Remember, they all began with an arbitrarily selected number. Although some readers might be motivated to investigate this further, others will just sit back in awe.

Here is an example of how this works with an arbitrarily selected starting number, 3618.

We select 3618.
The *largest* number formed with these digits is 8631.
The *smallest* number formed with these digits is 1368.
The difference is 7263.

Now using this number, 7263, we continue the process:
The largest number formed with these digits is 7632.
The smallest number formed with these digits is 2367.
The difference is 5265.

We repeat the process:
The largest number formed with these digits is 6552.
The smallest number formed with these digits is 2556.
The difference is 3996.

Again, we repeat the process:
The largest number formed with these digits is 9963.
The smallest number formed with these digits is 3699.
The difference is 6264.

Again, we repeat the process:
The largest number formed with these digits is 6642.
The smallest number formed with these digits is 2466.
The difference is 4176.

The largest number formed with these digits is 7641.
The smallest number formed with these digits is 1467.
The difference is 6174.

When one arrives at 6174, the number continuously reappears. Remember, all this began with an *arbitrarily selected* four-digit number. This then gets you into an endless loop by continuously returning to 6174. Your audience will be truly impressed with this trick!

A MULTIPLICATION TRICK FOR SPECIAL NUMBERS

Here is a trick for multiplying two-digit numbers whose units digit is 1, such as 21, 31, and 41. Follow along step-by-step and eventually the trick will become easier.

To multiply by 21: Double the number, then multiply by 10 and add the original number.

For example: to multiply 37 × 21, double 37 to yield 74, multiply by 10 to get 740, and then add the original number 37 to get 777.
Here's another example:

To multiply by 31: Triple the number, then multiply by 10 and add the original number.

For example: to multiply 43 × 31, triple 43 to yield 129, multiply by 10 to get 1290, and then add the original number 43 to get 1333.
For some more practice we offer the following:

To multiply by 41: Quadruple the number, then multiply by 10 and add the original number.

For example: to multiply 47×41, quadruple 47 to yield 188, multiply by 10 to get 1880, and then add the original number 47 to get 1927.
By now you should be able to recognize the pattern and apply the rule to other such two-digit numbers.

THE TRICK OF GUESSING THE DIVISOR

Here the trickster will use an interesting phenomenon in arithmetic to impress the audience. Ask the audience to choose any sequence of natural numbers that begins with 1 and ends before it reaches a prime number. Then have them multiply those numbers and add 1 to this product. This result is divisible by the prime number at which they stopped their sequence.

To show how this works, let's look at the first few such sequences that your audience might select. Consider the sequence that ends before the prime number 3 and perform the requested operation:

$1 \times 2 + 1 = 3$, which is divisible by 3.

Now consider the sequence that ends before the next prime number, 5:

$1 \times 2 \times 3 \times 4 + 1 = 25$, which is divisible by 5.

Consider the sequence that ends before the next prime number, 7:

$$1 \times 2 \times 3 \times 4 \times 5 \times 6 + 1 = 721, \text{which is divisible by } 7, \text{since } \frac{721}{7} = 103$$

Taking this one step further to the next prime number, 11:

$$1 \times 2 \times 3 \times 4 \times 5 \times 6 \times 7 \times 8 \times 9 \times 10 + 1 = 3,628,801, \text{which is divis-}$$

ible by 11, since $\dfrac{3,628,801}{11} = 329,891$.

To make this trick, which uses a mathematical curiosity, work as effectively as possible, you must think of a clever way to suit your audience's personality.

A CHALLENGE-QUESTION TRICK

Challenge your friend to find three numbers whose sum is 100, where the first number is divisible by 7, the second number by 17, and the third number by 27. If you begin with 17 and 27, whose sum is 44, then the number needed to get a sum of 100 is 56, which is divisible by 7. And there you have the requested answer: 56, 17, and 27.

ANOTHER CHALLENGE-QUESTION TRICK

Here is another challenge question to impress your friend. Ask her to find a six-digit number whose hundred thousands digit is 1 that, when multiplied by 3, produces another six-digit number, which is the same as the previous number, except that the 1 has moved from its original position to that of the units digit position.

This magic number is 142,857, which when multiplied by 3 gives us 428,571.

A TRICK TO SHOW YOUR TALENT

If you want to demonstrate your ability to read your friend's mind, try playing this trick on her. Ask your friend to select any two-digit number and square it. Then have her add 6 to this number and square that number as well. She now has two two-digit square numbers, which you will ask her to subtract. When she tells you her results, you will be able to tell her what her original numbers were.

To determine her original two numbers, take twice the original number 6 you asked her to add to her first-choice number to get 12. Then divide the number she gave you at the end by 12. That will give you a number halfway between her original two numbers. So, all you do now is take half of 6 and add and subtract it from the number you just obtained. This will give you her original two numbers.

Let's assume that your friend chooses 53 as a starting number. You then ask her to add 6 to that number, which gives her a second number, 59. Squaring each of the two numbers, she gets 2809 and 3481. When she subtracts these two numbers, it yields 762. This is the number she will reveal to you. All you do now is double the original 6 to get 12, then divide the 762 by 12 to get 56. This is the number midway between the two numbers that she selected. Take half of the original number we chose, 6, to get 3; then add 3 to 56 to get 59 and subtract 3 from 56 to get 53 to get your friend's two numbers, 53 and 59.

This trick could also be done using a number other than 6, using the same basic procedure. Once again, simple algebra will help understand why this trick works.

GUESS THE DICE TRICK

Imagine that, when your friend tosses a pair of dice, you can tell what numbers are showing without seeing them. This trick will enable you to do that. Have your friend toss a pair of dice without showing you the result. Then have your friend write the number that is on one of the dice, multiply that number by 2, and then add 1. Then have him multiply that result by 5 and add the numbers showing on the second die. This last number is the one that your friend is to share with you.

Now subtract 5 from the number he shares with you. The tens digit of the result will be what one die showed, and the units digit will be what the other die showed.

We now illustrate this trick for a specific case. Supposing the pair of dice show 3 and 5. Following the process described above, your friend writes down one of the numbers shown, 3. He then doubles the 3 to get 6 and adds 1 to get 7. When he multiplies that by 5, he will get 35. His last calculation is to take the number from the second die, which, in this case, was 5, and add it to the previously found number, 35, to get 40. This number is then presented to the trickster, who then merely subtracts 5 to get 35. Thus, the trickster can conclude that these two digits are the numbers shown originally on the two dice, 3 and 5.

A CLOCK TRICK

This magic clock trick is quite entertaining and open to some speculation on why it works. Consider a normal clock, with the usual numerals from 1 to 12.

We now place a 3 at the usual 1 position, and then place multiples of 3 at each position (clockwise) until we get to the 12 position. Then besides these multiples of 3, we begin with a 4 at the usual 1 position and similarly continue with multiples of 4 until we get to the 12 position again. This is shown in Figure 1.29.

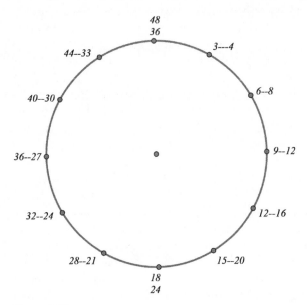

Figure 1.29

We now present some unusual results from this arrangement of numbers. First, by subtracting each pair of numbers, you get the hour position at each place. For example, the 5-hour position can be obtained by subtracting $20 - 15 = 5$. If you take the difference of each pair of numbers and add that to the larger of the two numbers, you will get the number of minutes to that position. For example, the 25-minute position can be obtained by subtracting $20 - 15 = 5$ and adding it to the larger of the two numbers, $20 + 5 = 25$.

Moreover, taking the square root of the sum of the squares of each pair of numbers gives you the number of minutes represented at that position. We can see that by calculating $20^2 + 15^2 = 625 = 25^2$. You might like to try this on some of the other clock positions.

BECOME A MENTAL CALCULATOR[3]

Before modern electronic calculators and computers, all mathematical calculations had to be done using paper and pencil. Slide rules or more sophisticated mechanical devices were the only technological assistance.

It is hard to imagine how today's world would look if that were still the case. Did you know that "computer" originally referred to a person who carried out calculations or computations? The word retained this meaning until the middle of the twentieth century. Its meaning gradually changed as technological advances led to more and more efficient machines, eventually causing the extinction of "human computers." Nowadays prodigies in mental calculation appear more often in television shows than in research centers. One such mental-calculation expert is Dr. Arthur Benjamin, a professor of mathematics at Harvey Mudd College in California. At the turn of the twentieth century, however, research agencies like the National Advisory Committee for Aeronautics (NACA, founded in 1915, transferred to NASA in 1958) still relied on human computers and, naturally, they scouted the highest-performance individuals in this profession. In particular, mental calculators were in great demand. Although there is no comparable job profile today, there are still people around who practice mental calculation very intensively. Every two years, the world's best mental calculators compete for the Mental Calculation World Cup, first held in 2004 in Germany. A common misbelief is that mathematicians are particularly good in mental calculation. But in fact, many mathematicians like to emphasize that they have problems with mental arithmetic. Such statements often come with a pinch of coquetry, but there is some truth behind them. A mathematician does not have to be good in mental arithmetic. People with limited mathematical education or interest often think that mathematics is "all about computing numbers," which is not at all the case. Being an excellent mathematician but lousy in mental arithmetic is actually no contradiction. Of course, some outstanding mathematicians are or were also exceptional mental calculators. The Hungarian-American mathematician John von Neumann (1903–1957), one of the greatest mathematicians of the twentieth century, was a child prodigy in language, memorization, and mathematics. At the age of six, he could divide one eight-digit number by another in his head. He also possessed an amazing memory. A brief glance at a page of the telephone directory sufficed for him to memorize all its names and numbers.

In this section we show you how to impress your friends with seemingly incredible mental calculation skills! Ask them to secretly choose any number between 1 and 100 and take its fifth power using a pocket calculator. Then ask them to tell you the result. The trick is that you will be able to almost instantly reveal the original number. For instance, your friend may tell you 69,343,957. After a few seconds of pretended intense thinking you will have calculated its fifth root, which is 37. Terrific, isn't it? Of course, there is a trick behind this mind magic. It comprises just a few mathematical observations and—this is the harder part—memorizing the magnitude of a couple of numbers. However, anyone can do it after half an hour of training. So how does it work?

As you have probably already expected, you don't have to memorize all 100 possible outcomes when someone takes the fifth power of a number between 1 and 100. Truth be told, though, you still have to memorize some of those numbers, or at least their magnitudes. First we present the crucial observations that make the trick work. Note that $1^5 = 1$ and $100^5 = \left(10^2\right)^5 = 10^{10} = 10,000,000,000$, which are the two easy cases. They won't be chosen anyway, if your friends really want to challenge you. All numbers from 1 to 99 can be written as two-digit numbers D_1D_2, where D_1 is the first digit (or zero) and D_2 is the last digit of the number. For example, in the case of 37, we have $D_1 = 3$ and $D_2 = 7$. Thus, we can write any two-digit number as $10D_1 + D_2$, where D_1 and D_2 are one-digit numbers. Now comes the first important observation: Calculating the fifth power of a two-digit number yields:

$$\left(10D_1 + D_2\right)^5 = \left(\mathbf{1} \times 10^5 \times D_1^{\,5}\right) + \left(\mathbf{5} \times 10^4 \times D_1^{\,4}D_2\right) + \left(\mathbf{10} \times 10^3 \times D_1^{\,3}D_2^{\,2}\right) +$$

$$\left(\mathbf{10} \times 10^2 \times D_1^{\,2}D_2^{\,3}\right) + \left(\mathbf{5} \times 10 \times D_1D_2^{\,4} + \mathbf{1} \times D_2^{\,5}\right).$$

The bold numbers as well as the exponents can easily be determined with the help of Pascal's triangle. All summands on the right-hand side except the last one contain a factor 10. This implies that the last digit of the number $\left(10D_1 + D_2\right)^5$ is identical to the last digit of the number $D_2^{\,5}$. But that's not all! In Figure 1.30 we have listed the fifth powers of all one-digit numbers.

D_2	0	1	2	3	4	5	6	7	8	9
$D_2^{\,5}$	0	1	32	243	1024	3125	7776	16807	32768	59049

Figure 1.30

For any one-digit number, the last digit of its fifth power is just the original number. Combining this observation with the first one, we conclude that the last digit of the fifth power of any two-digit number coincides with the last digit of the original number. Hence, if you are told the fifth power of a two-digit number, you will immediately know the last digit of the number your friend chose. Returning to our example, $37^5 = 69,343,957$, so the 7 is easily determined.

But how do we get the first digit of the original number? Well, this requires a little more work. Basically, we have to find out whether $69,343,957$ lies between 10^5 and 20^5, or between 20^5 and 30^5, or between 30^5 and 40^5, and so forth. Note that these "threshold numbers" are simply given by 10^5 times the numbers shown in Figure 1.30 (since $30^5 = 3^5 \cdot 10^5$, and analogously for the other multiples of 10). Therefore, we can ignore the last five digits of $69,343,957$ to figure out the first digit of the original number. Doing so, we are left with 693, which lies between 243 and 1024 in the second row of Figure 1.30. Hence, the first digit of the original number must have been 3, and you can triumphantly claim that the fifth root of 69,343,957 should be 37.

To determine a two-digit number from its fifth power, you just have to memorize the numbers in Figure 1.30. Let us play through another example to see how the magic trick works. Suppose your friend calls out 4,182,119,424. You write it down (or let her or him write it) and note the last digit, which is 4. This will be the second digit of the original number. Then you ignore the last five digits of the number presented to you and concentrate on the remaining part, 41,821. Having memorized the numbers in figure 1.30, it is now easy to conclude that the first digit must have been 8, since 41,821 is between the two numbers 32,768 and 59,049 in the table. Therefore, the original number must have been 84. In fact, you don't even have to memorize all the numbers in Figure 1.30 exactly. It suffices to memorize the properly rounded approximations. Such easily memorizable threshold numbers are shown in Figure 1.31. Here, N_{-5} denotes the number that remains after dropping the last five digits of the fifth power of a given two-digit number $10D_1 + D_2$ (or after shifting the decimal point five positions to the left, respectively).

$D_1 = 0$	$N_{-5} < 1$
$D_1 = 1$	$1 \leq N_{-5} < 30$
$D_1 = 2$	$30 < N_{-5} < 230$
$D_1 = 3$	$230 < N_{-5} < 1000$
$D_1 = 4$	$1000 < N_{-5} < 3000$
$D_1 = 5$	$3000 < N_{-5} < 7500$
$D_1 = 6$	$7500 < N_{-5} < 16000$
$D_1 = 7$	$16000 < N_{-5} < 32000$
$D_1 = 8$	$32000 < N_{-5} < 56000$
$D_1 = 9$	$56000 < N_{-5}$

Figure 1.31

Knowing all the numbers in Figure 1.31 from memory is sufficient to work out the original number when somebody tells you the fifth power of any two-digit number. A tough but neat trick!

THE EXTENDED MAGIC SQUARE

Magic squares have been around for centuries. Perhaps the most famous one is pictured at the upper right in the famous German artist Albrecht Dürer's etching *Melencolia I,* shown in Figure 1.32. For any magic square, the sums of the numbers in any row, column or diagonal are the same—in this case the sums are all 34.

Figure 1.32

For this trick, however, we will use a much simpler 3×3 magic square, shown in Figure 1.33. In this 3×3 magic square the sum of the rows, columns, and diagonals is 15.

The trick here is to ask your audience if they can create a magic square where the *products* of the numbers in each of the rows, columns, and diagonals, rather than their *sums*, are the same.

We need simply to change each of the numbers in the cells to a power determined by the original numbers in the previous magic square cells to the same base. We will choose base 2, since it is the easiest one to work with. Notice that the exponents in Figure 1.34 correspond to the cell entries in Figure 1.33.

The final magic square that shows multiplication instead of addition is shown in Figure 1.35.

2	7	6
9	5	1
4	3	8

Figure 1.33

2^2	2^7	2^6
2^9	2^5	2^1
2^4	2^3	2^8

Figure 1.34

4	128	64
512	32	2
16	8	256

Figure 1.35

The product of each of the rows, columns, and diagonals is 32,768. Note that we could have used any other base besides 2 and still constructed a magic square where the *products* of the rows, columns, and diagonals are the same.

GENERATING PALINDROMES

A palindrome is a number that reads the same in both directions, such as 1331, or 555, or 12,321. The point here is not only to show how you can generate a palindrome from a two-digit number but also for you, as the trickster, to know how many steps it would take to reach a palindrome using this technique. To generate a palindrome, the procedure is very simple. Suppose you begin with a two-digit number, such as 25. When you add this to its reversal, which is 52, you get 77, which is a palindrome. On the other hand, if you start off with 68 and add it to its reversal, 86, you get 154. You then need to repeat this procedure, 154 + 451 = 605, and once again repeat the procedure, 605 + 506 = 1111, which again is a palindrome. So, a trickster can use this technique in a variety of ways. However, the trickster will be able to embellish this technique by knowing how many steps are required to reach a palindrome when beginning with any two-digit number. When the sum of the digits of a two-digit number is less than 10, only one step will be required. On the other hand, if the sum of the digits of the initial two-digit number is 10, 11, 12, 13, 14, 15, or 16, the number of reversals will be 2, 1, 2, 2, 3, 4, and 6, respectively. When the sum of the digits of the initial two-digit number is 17, such as for 89 or 98, more than 10 reversals will be required. This extra insight into this process will give the trickster various ways of approaching the challenge to his audience. It should be noted that there are certain numbers where continuous reversal additions will not lead to a palindromic number. One such number is 196.

REPRESENT NUMBERS USING ONLY THE NUMERAL 4

The trickster will probably want a crib sheet as he performs the following trick. The trickster should ask the audience to select any number from 1 to 100 and then show how that number can be represented using only the numeral 4. Before giving the response, the trickster might want to let the audience try it first. Following is a list of possible ways in which these numbers can be expressed in terms of the numeral 4.

$$0 = 44 - 44 = \frac{4}{4} - \frac{4}{4}$$

$$1 = \frac{4+4}{4+4} = \frac{\sqrt{44}}{\sqrt{44}} = \frac{4+4-4}{4}$$

$$2 = \frac{4 \times 4}{4+4} = \frac{4-4}{4} + \sqrt{4} = \frac{4}{4} + \frac{4}{4}$$

$$3 = \frac{4+4+4}{4} = \sqrt{4} + \sqrt{4} - \frac{4}{4} = \frac{4 \times 4 - 4}{4} = 4 - 4^{4-4}$$

$$4 = \frac{4-4}{4} + 4 = \frac{\sqrt{4 \times 4} \times 4}{4} = (4-4) \times 4 + 4$$

$$5 = \frac{4 \times 4 + 4}{4}$$

$$6 = \frac{4+4}{4} + 4 = \frac{4\sqrt{4}}{4} + 4$$

$$7 = \frac{44}{4} - 4 = \sqrt{4} + 4 + \frac{4}{4} = (4+4) - \frac{4}{4}$$

$$8 = 4 \times 4 - 4 - 4 = \frac{4(4+4)}{4} = 4 + 4 + 4 - 4$$

$$9 = \frac{44}{4} - \sqrt{4} = 4\sqrt{4} + \frac{4}{4} = \frac{4}{4} + 4 + 4$$

$$10 = 4 + 4 + 4 - \sqrt{4} = \frac{44 - 4}{4}$$

$$11 = \frac{4}{4} + \frac{4}{.4} = \frac{44}{\sqrt{4} + \sqrt{4}}$$

$$12 = \frac{4 \times 4}{\sqrt{4}} + 4 = 4 \times 4 - \sqrt{4} - \sqrt{4} = \frac{44 + 4}{4}$$

$$13 = \frac{44}{4} + \sqrt{4}$$

$$14 = 4 \times 4 - 4 + \sqrt{4} = 4 + 4 + 4 + \sqrt{4} = \frac{4!}{4+4+4} = 4! - (4 + 4 + \sqrt{4})$$

$$15 = \frac{44}{4} + 4 = \frac{\sqrt{4} + \sqrt{4} + \sqrt{4}}{.4}$$

$$16 = 4 \times 4 - 4 + 4 = \frac{4 \times 4 \times 4}{4} = 4 + 4 + 4 + 4$$

$$17 = 4 \times 4 + \frac{4}{4}$$

$$18 = \frac{44}{\sqrt{4}} - 4 = 4 \times 4 + 4 - \sqrt{4} = 4 \times 4 + \frac{4}{\sqrt{4}} = \frac{4! + 4! + 4!}{4}$$

$$19 = \frac{4 + \sqrt{4}}{.4} + 4 = 4! - 4 - \frac{4}{4}$$

$$20 = 4 \times 4 + \sqrt{4} + \sqrt{4} = \left(4 + \frac{4}{4}\right) \times 4$$

$$21 = 4! - 4 + \frac{4}{4}$$

$$21 = 4! 4 \times 4 + 4 + \sqrt{4}$$

$$22 = \frac{4}{4}(4!) - \sqrt{4} = 4! - \frac{((4+4)/4)}{4} = \frac{44}{4} \times \sqrt{4} = -4 + \frac{4}{4}$$

$$23 = 4! - \sqrt{4} + \frac{4}{4} = 4! - 4^{4-4}$$

$$24 = 4 \cdot 4 + 4 + 4$$

$$25 = 4! - \sqrt{4} + \frac{4}{4} = 4! + \sqrt{(4 + 4 - 4)} = \left(4 + \frac{4}{4}\right)^{\sqrt{4}}$$

$$26 = \frac{4}{4}(4!) + \sqrt{4} = 4! + \sqrt{4 + 4 - 4} = 4 + \frac{44}{\sqrt{4}}$$

$$27 = 4! + 4 - \frac{4}{4}$$

$$28 = (4 + 4) \times 4 - 4 = 44 - 4 \times 4$$

$$29 = 4! + 4 + \frac{4}{4}$$

$$30 = 4! + 4 + 4 - \sqrt{4}$$

$$31 = \frac{((4 + \sqrt{4})! + 4!)}{4!}$$

$$32 = (4 \times 4) + (4 \times 4)$$

$$33 = 4! + 4 + \frac{\sqrt{4}}{.4}$$

$$34 = \left(4 \times 4 \times \sqrt{4}\right) + \sqrt{4} = 4! + \left(\frac{4!}{4}\right) + 4 = \sqrt{(4^4)} \times \sqrt{4} + \sqrt{4}$$

$$35 = 4! + \frac{44}{4}$$

$$36 = (4 + 4) \times 4 + 4 = 44 - 4 - 4$$

$$37 = 4! + \frac{\left(4! + \sqrt{4}\right)}{\sqrt{4}}$$

$$38 = 44 - \frac{4!}{4}$$

$$39 = 4! + \frac{4!}{4 \times .4}$$

$$40 = (4! - 4) + (4! - 4) = 4 \times \left(4 + 4 + \sqrt{4}\right)$$

$$41 = \frac{4! + \sqrt{4}}{.4} - 4!$$

$$42 = 44 - 4 + \sqrt{4} = (4! + 4!) - \frac{4!}{4}$$

$$43 = 44 - \left(\frac{4}{4}\right)$$

$$44 = 44 + 4 - 4$$

$$45 = 44 + \frac{4}{4}$$

$$46 = 44 + 4 - \sqrt{4} = (4! + 4!) - \left(\frac{4}{\sqrt{4}}\right)$$

$$47 = (4! + 4!) - \frac{4}{4}$$

$$48 = (4 \times 4 - 4) \times 4 = 4 \times (4 + 4 + 4)$$

$$49 = (4! + 4!) + \frac{4}{4}$$

$$50 = 44 + \left(\frac{4!}{4}\right) = 44 + 4 + \sqrt{4}$$

$$51 = \frac{(4!- 4 + .4)}{.4}$$

$$52 = 44 + 4 + 4$$

$$53 = 4! + 4! + \frac{\sqrt{4}}{.4}$$

$$54 = 4! + 4! + 4 + \sqrt{4}$$

$$55 = \frac{(4! - 4 + \sqrt{4})}{.4}$$

$$56 = 4! + 4! + 4 + 4 = 4 \times (4 \times 4 - \sqrt{4})$$

$$57 = \left(\frac{4!- \sqrt{4}}{.4}\right) + \sqrt{4}$$

$$58 = ((4! + 4) \times \sqrt{4}) + \sqrt{4} = 4! + 4! + \frac{4}{.4}$$

$$59 = \frac{\left(4! - \sqrt{4}\right)}{.4} + 4 = \frac{4!}{.4} - \frac{4}{4}$$

$$60 = 4 \times 4 \times 4 - 4 = \frac{4^4}{4} - 4 = 44 + 4 \times 4$$

$$61 = \frac{(4! + \sqrt{4})}{.4} - 4 = \frac{4!}{.4} + \frac{4}{4}$$

$$62 = 4 \times 4 \times 4 - \sqrt{4}$$

$$63 = \frac{(4^4 - 4)}{4}$$

$$64 = 4\sqrt{4} \times 4\sqrt{4} = 4 \times (4! - 4 - 4) = (4 + 4) \times (4 + 4)$$

$$65 = \frac{(4^4 + 4)}{4}$$

$$66 = 4 \times 4 \times 4 + \sqrt{4}$$

$$67 = \frac{4! + \sqrt{4}}{.4} + \sqrt{4}$$

$$68 = 4 \times 4 \times 4 + 4 = \frac{4^4}{4} + 4$$

$$69 = \left(\frac{4! + \sqrt{4}}{.4}\right) + 4$$

$$70 = \frac{(4 + 4)!}{4! \times 4!} = 44 + 4! + \sqrt{4}$$

$$71 = \frac{4! + 4.4}{.4}$$

$$72 = 44 + 4! + 4 = 4 \times \left(4 \times 4 + \sqrt{4}\right)$$

$$73 = \frac{4! \cdot \sqrt{4} + \sqrt{.4}}{\sqrt{.4}}$$

$$74 = 4! + 4! + 4! + \sqrt{4}$$

$$75 = \frac{4! + 4 + \sqrt{4}}{.4}$$

$$76 = (4! + 4! + 4!) + 4$$

$$77 = \left(\frac{4}{.4}\right)^{\sqrt{4}} - 4$$

$$78 = 4 \times (4! - 4) - \sqrt{4}$$

$$79 = 4! + \frac{4! - \sqrt{4}}{.4}$$

$$80 = (4 \times 4 + 4) \times 4$$

$$81 = \left(4 - \left(\frac{4}{4}\right)\right)^4 = \left(\frac{4!}{4\sqrt{4}}\right)^4$$

$$82 = 4 \times (4! - 4) + \sqrt{4}$$

$$83 = \frac{4! - .4}{.4} + 4!$$

$$84 = 44 \times \sqrt{4} - 4$$

$$85 = \frac{\frac{4! + 4}{.4}}{.4}$$

$$86 = 44 \times \sqrt{4} - \sqrt{4}$$

$$87 = 4 \times 4! - \frac{4}{.4} = 44\sqrt{4} - i^4$$

$$88 = 4 \times 4! - 4 - 4 = 44 + 44$$

$$89 = 4! + \frac{4! + \sqrt{4}}{.4}$$

$$90 = 4 \times 4! - 4 - \sqrt{4} = 44 \times \sqrt{4} + \sqrt{4}$$

$$91 = 4 \times 4! - \frac{-\sqrt{4}}{.4}$$

$$92 = 4 \times 4! - \sqrt{4} - \sqrt{4} = 44 \times \sqrt{4} + 4$$

$$93 = 4 \times 4! - \frac{4}{.4}$$

$$94 = 4 \times 4! + \sqrt{4} - 4$$

$$95 = 4 \times 4! - \frac{4}{4}$$

$$96 = 4 \times 4! + 4 - 4 = 4! + 4! + 4! + 4!$$

$$97 = 4 \times 4! + \frac{4}{4}$$

$$98 = 4 \times 4! + 4 - \sqrt{4}$$

$$99 = \frac{44}{.44} = 4 \times 4! + \frac{\sqrt{4}}{\sqrt{.04}} = \frac{4}{4\%} - \frac{4}{4}$$

$$100 = 4 \times 4! + \sqrt{4} + \sqrt{4} = \left(\frac{4}{.4}\right) \times \left(\frac{4}{.4}\right) = \frac{44}{.44}$$

An ambitious audience might want to extend this was beyond 100.

NOTES

1. For a proof that this relationship holds as started, see Ross Honsberger, *Ingenuity in Mathematics* (New York: Random House, 1970), 147–56.

2. If you know about permutations, you will recognize that these six numbers are the *only* ways to write a number with three different digits. That is why we ask for three *distinct* digits in the first place!

3. Contributed by Dr. Christian Spreitzer

2

Unexpected Geometry Tricks

The tricks in geometry are different from those in the previous chapter, which dealt largely with arithmetic surprises and were often justified through algebraic techniques. In geometry the trickster will present situations that will appear as though the trick was performed because of the surprising results to which the trickster's friend was guided. In some cases, the friend will go a few steps in one direction and surprisingly end up with what superficially is inexplicable—an unexpected shape of polygon, lines that turn out to be concurrent, points that surprisingly are collinear, or a number of points lying on the same circle. There are also tricks that show that things we logically feel are impossible are in fact true. These surprising results will give the impression that a trick has been played, while in actual fact, it is the power of mathematics that allows them to occur. We begin this chapter with visual tricks that will make the reader a much more critical observer in the future. Join us now as we investigate the surprising mathematical phenomena that the trickster will guide his friend to appreciate. We begin with visual deceptions.

OPTICAL TRICKS

We can begin by comparing the two segments shown in Figure 2.1. Which of the two segments looks longer? Most people say that the one on the right looks longer. We tricked you! It is not; the lines are the same length. In

Figure 2.1

Figure 2.2 you might think that the bottom segment also looks longer. In actuality, the segments have the same length. Fooled again![1]

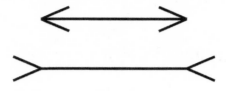

Figure 2.2

Here are some more optical tricks. In Figure 2.3 the crosshatched segment appears longer than the clear one. In Figure 2.4 on the right side, the narrower and vertical stick appears to be longer than the other two, even though on the left they are shown to be the same length.

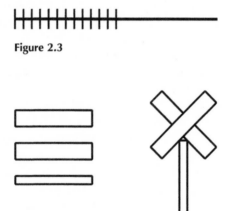

Figure 2.3

Figure 2.4

Here is an often-seen trick, which we can also refer to as an optical illusion. Figure 2.5 very cleverly tricks us to believe that AB is longer than BC. This isn't true, since $AB = BC$.

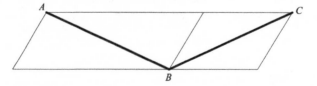

Figure 2.5

In Figure 2.6 the vertical segment clearly, again, tricks us to imagine that it is longer than the horizontal segment, but it isn't. The lines are equal in length. When we admire the two curves in Figure 2.7, the top curve seems broader than the lower curve. Here we have been duped, because the two curves are congruent.

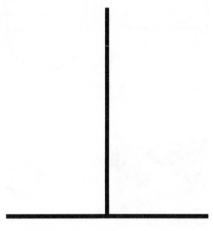

Figure 2.6

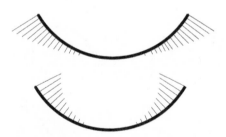

Figure 2.7

The square between the two semicircles in Figure 2.8 looks bigger than the square to the left, but the two squares are the same size. Once again, we have been fooled! In Figure 2.9 the square within the large black square looks smaller than that to the right; but, again, we have been tricked because this is an optical illusion, as they are the same size.

We see further evidence of tricking the senses in Figure 2.10, where the larger circle inscribed in the square (on the left) appears to be smaller than the circle circumscribed about the square (on the right). No, the circles are the same size!

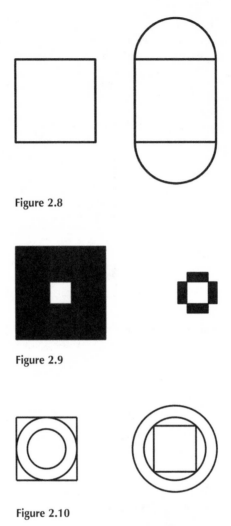

Figure 2.8

Figure 2.9

Figure 2.10

Figures 2.11, 2.12, and 2.13 show how relative placement can trick us about the appearance of a geometric diagram. In Figure 2.11, the center square appears to be the largest of the group, but it isn't; all the squares are equal in size. A rather upsetting trick is seen in Figure 2.12, where the black center circle on the left appears to be smaller than the black center circle on the right. Again, it is not, as they are equal in size.

Our final optical trick is shown in Figure 2.13, where the center sector on the left appears to be smaller than the center sector on the right. Once again, we have been tricked into judging sizes incorrectly. In all of these cases, the two figures that appear not to be the same size are, in fact, the same size!

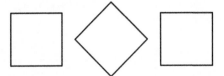

Figure 2.11

Figure 2.12

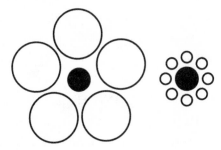

Figure 2.13

A VISUAL TRICK

Up until now, the optical illusions were somewhat enlightening and entertaining. However, we now present a challenge. The diagram in Figure 2.14 shows three different possible arrangements of cubes. The trick is to identify three situations. You should be able to see a small cube inside a larger cube; a small cube cut out of a larger cube; and a third one, which is a bit more difficult to imagine—a small cube placed externally at the vertex of the larger cube. If you get all three, you should win a prize. This can be a fun activity to present to your audience.

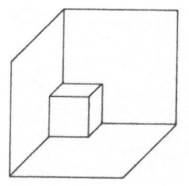

Figure 2.14

THE OVERLAPPING AREAS TRICK

It is nice to impress others with tricks that seem to require a lot of computation and can be done quickly. One such trick involves two overlapping circles, as shown in Figure 2.15. Here the trickster asks the audience to find very quickly the difference of the areas of the nonoverlapping regions of the two circles.

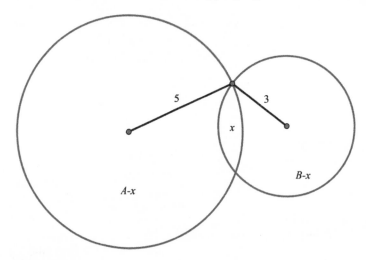

Figure 2.15

We begin with two circles of areas A and B. When these two areas overlap, we can call the overlapping area x, which then allows us to refer to the nonoverlapping regions' areas as $A - x$ and $B - x$. When we take the difference of these areas—normally the larger minus the smaller—we find the difference of the areas to be $A - B$. Hence, the difference of the areas of the nonoverlapping regions is simply the difference of the areas of the circles, namely, $25\pi - 9\pi = 16\pi$. The solution to this question is surprisingly simple, yet counterintuitive—however, we will have the trickster properly prepared.

THE SEWER COVER CONUNDRUM

Have you ever wondered why sewer covers are always round? Well, as a trickster you might want to let your audience ponder this question. As you can see in Figure 2.16, the answer is: a circular cover cannot fall into the hole.

Figure 2.16

You should know, however, that other shapes can be used that also will not fall into the hole. One of these is the Reuleaux triangle, named after the German engineer Franz Reuleaux (1829–1905), and is shown in Figure 2.17.

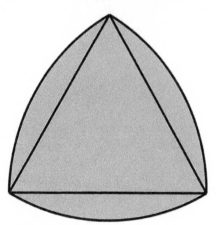

Figure 2.17

The Reuleaux triangle is formed by constructing an equilateral triangle and drawing a circular arc on each side with the center of the arc's circle at the opposite vertex.

THE TRICK TO CONSTRUCT THE UNEXPECTED
PARALLELOGRAM

There are some geometric tricks so simple that they can be drawn on a paper napkin in a restaurant. Suppose you would like to impress someone with a trick that will get them wondering why they have never seen anything like it before. Begin by having your audience draw a quadrilateral with no equal sides or equal angles. In other words, have them draw an ugly quadrilateral. It is always best to have the audience draw a variety of shaped quadrilaterals to show that no secret is being hidden. Have them locate the midpoint of each side of their quadrilateral and then connect these midpoints consecutively. Drawn properly, the resulting shape in each case should be a parallelogram, as you can see with our example in Figure 2.18. They should be amazed that no matter what shape quadrilateral they began with, the end result will always be a parallelogram inside the original quadrilateral.

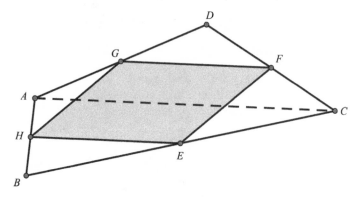

Figure 2.18

To justify this result we merely apply a theorem that is part of the usual high school curriculum. In any triangle, a line segment joining midpoints of two sides of the triangle is parallel to the third side of the triangle and one half its length. If we draw one diagonal, say *AC*, in this original quadrilateral, we will have formed two triangles (*ADC* and *ABC*). In triangle *ABC* we then have *HE* parallel to and one half the length of *AC*; and in triangle *ADC* we have *GF* parallel to and one half the length of *AC*. Therefore, we have *GF* parallel to *HD*, and *GF* is equal to *HE*. Then we can show that the quadrilateral formed by joining consecutive midpoints of the quadrilateral will have two sides (*GF* and *HE*) equal in length and parallel, since they are both parallel and half the length of *AC*. This makes the newly formed quadrilateral a parallelogram.

THE TRICK TO CONSTRUCT ANOTHER UNEXPECTED PARALLELOGRAM

Along the same line of thinking, here is another trick for an audience that is impressed with surprises. Once again, begin by having them draw an "ugly" quadrilateral (randomly drawn and with no sides or angles equal), like the randomly drawn quadrilateral *ABCD* along with its diagonals shown in Figure 2.19. Then have them locate the midpoints of a pair of opposite sides, such as we did with sides *AB* and *DC*, whose midpoints are marked as points *F* and *E*, respectively. However, this time we locate the midpoints of the diagonals at points *M* and *N*. Your audience will be amazed that they once again have constructed a parallelogram, *ENFM*. The rationale, or justification, follows the same path as the previous illustration in Figure 2.18.

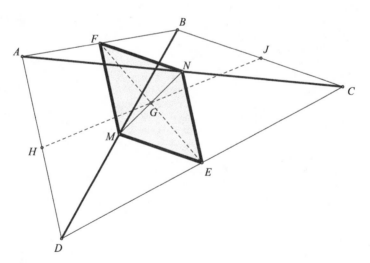

Figure 2.19

Moreover, as a "bonus," you can further impress your audience by having them draw the line joining midpoints of the other two sides of the quadrilateral *ABCD*. They will see that the point of intersection of the diagonals of every such parallelogram is the same as that of the lines joining the midpoints of the opposite sides of the original quadrilateral. Remember to do this stepwise so that each move, or shall we say trick, will leave them somewhat amazed. Stress the beauty of the concurrencies that exist here.

THE TRICK TO CONSTRUCT THE UNEXPECTED RHOMBUS

We can extend the technique, with which we are now comfortable, so as to guide our audience to the trick way of constructing a rhombus. To do this, ask your audience to construct any "ugly" quadrilateral, yet whose diagonals are equal, such as quadrilateral *ABCD* shown in Figure 2.20, where diagonals $AC = BD$. By connecting the midpoints of the four sides of the quadrilateral, we find that the newly formed quadrilateral *EFGH* is, in fact, a rhombus.

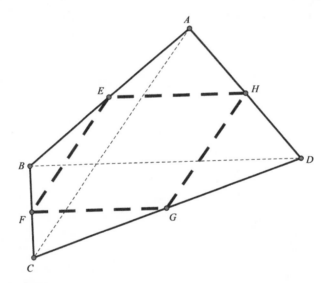

Figure 2.20

Explain this to your audience by simply referring back to the previous general parallelogram trick, where both pairs of opposite sides are one half the length of each of the equal diagonals, and therefore, all the sides are equal.

THE TRICK TO CONSTRUCT THE UNEXPECTED
RECTANGLE

Have the audience draw another ugly quadrilateral, using the same technique as above. However, there is one proviso: the diagonals should be perpendicular, as we can see in Figure 2.21.

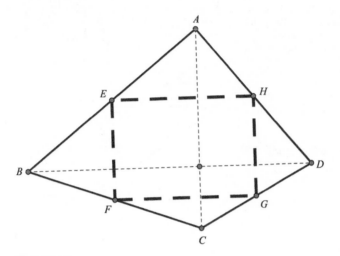

Figure 2.21

Since the sides of the newly formed quadrilateral are parallel to the two diagonals, and the diagonals are perpendicular, the adjacent sides of the newly formed quadrilateral are perpendicular, resulting in a rectangle.

THE TRICK TO CONSTRUCT THE UNEXPECTED SQUARE

As we develop tricks based on joining the midpoints of a quadrilateral, the logical next trick would be to set up the analogous construction of a square. You ask your audience to begin with another ugly quadrilateral, but this time one where the diagonals are perpendicular and the same length. The newly formed quadrilateral will be a square. We see this in Figure 2.22.

Not only will you be varying your trick platform, but in this particular case you also will be refreshing your audience's knowledge of high school geometry.

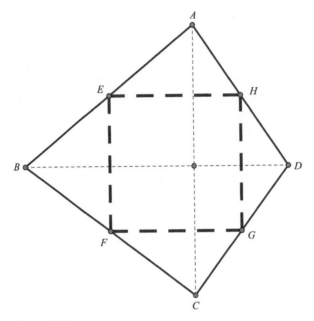

Figure 2.22

THE QUADRILATERAL MIDPOINTS TRICK

With this trick you ask your friend to locate the midpoints of each of the four sides of a *cyclic quadrilateral*, whose four vertices lie on the same circle. In addition to that, your friend is to locate the midpoints of the two diagonals. Your friend will find that you have presented a trick situation in which the connections of the midpoints are concurrent. Let's take a look at Figure 2.23, where quadrilateral *ABCD* is inscribed in circle *O*, and the midpoints of the sides of the quadrilateral are points *E, F, G,* and *H*. The midpoints of the two diagonals are denoted by points *M* and *N*. When your friend joins the midpoints of the opposite sides of the quadrilateral, they will meet at point *P*. Then when your friend joins the midpoints of the two diagonals amazement will set in, noticing that this line contains point *P* as well.

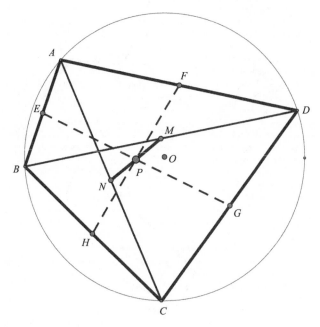

Figure 2.23

THE SURPRISING APPEARANCE OF A RECTANGLE TRICK

Some tricks in geometry often require you to use construction tools, or at least have the ability to draw geometric shapes and bisect lines and angles without tools. One such trick is to begin with a circle and then choose any four points on the circle that will give some form of ugly quadrilateral—that is, a quadrilateral with no equal sides or equal angles and no parallel sides. Let us assume that your friend's quadrilateral *ABCD* inscribed in circle *O* looks like the one shown in Figure 2.24.

Your friend is then asked to bisect each of the angles of the quadrilateral just constructed and extend the angle bisectors to reach the opposite side of the circle. Regardless of the shape of the original quadrilateral *ABCD*, the four points of contact of the angle bisectors on the circle, when joined consecutively, will yield a rectangle. In other words, this geometric trick has produced rectangle *EFGH*. Quite surprising!

The trickster may want to take this one step further and show their friend that points *W, X, Y,* and *Z* (shown in Figure 2.25) will also lie on a unique circle. This is certainly uncommon, since, while three noncollinear points always lie on a unique circle, having more than three points all on the same circle is

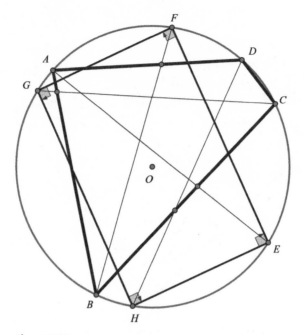

Figure 2.24

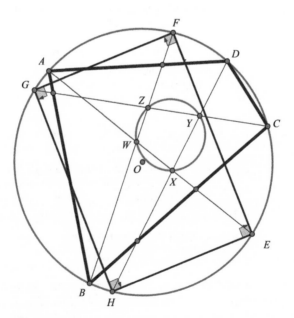

Figure 2.25

quite noteworthy. This is just a little bonus on the initial trick. Consequently, you have a double trick that should be carefully explained to the audience so that they can appreciate the uniqueness of this result.

THE TRICK OF UNUSUAL QUADRILATERAL ANGLE BISECTORS

Once again, we consider an ugly quadrilateral. In this case, however, it will have one special property: the angle bisectors of one pair of opposite angles intersect on one of the diagonals of the quadrilateral. As trickster, you might begin with a quadrilateral that looks like the one shown in Figure 2.26, where one pair of angle bisectors, *AP* and *CP*, meet on diagonal *BD*. You show this to your friend and tell him that you, as trickster, when he draws the other pair of angle bisectors, can predict where they will intersect. He will be quite surprised to find that bisectors *DQ* and *BQ* also intersect on the other diagonal, *AC*.

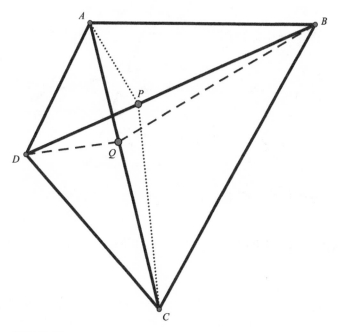

Figure 2.26

THE SURPRISING TRICK OF THE ANGLE BISECTORS OF A QUADRILATERAL

Begin by having your friend draw a quadrilateral of any shape, which is technically called a general quadrilateral or, as we have called it, an ugly quadrilateral. Now, have your friend construct the bisectors of each of the four angles of their self-constructed quadrilateral and extend them until each bisector intersects the bisectors of the adjacent angles of the quadrilateral. In Figure 2.27, we show the general quadrilateral *ABCD* with each of the four angle bisectors drawn so that they meet adjacent angle bisectors. The trick here is to notice that the four points of intersection that your friend will have established, *E*, *F*, *G*, and *J*, all lie on the same circle. Alert your friend about the wonder of this finding. Any three noncollinear points will always lie on a unique circle; however, four noncollinear points lie on the same circle only in special cases and are known as concyclic points—a rather special arrangement.

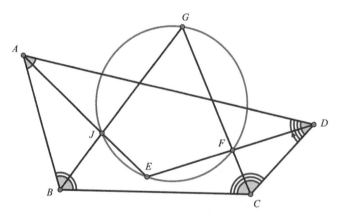

Figure 2.27

THE ANGLE BISECTORS OF A PARALLELOGRAM TRICK

A trickster can take the previous trick one step further. Have your friend begin with a parallelogram rather than a general quadrilateral, then draw the four angle bisectors as before. Here your friend will create a rectangle, which, of course, is formed by the four concyclic points *E*, *F*, *G*, and *J*, as shown in Figure 2.28.

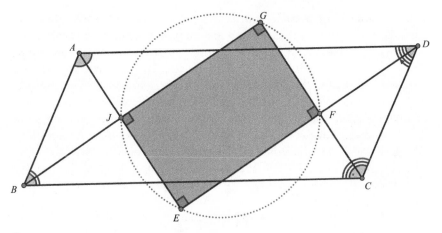

Figure 2.28

THE ANGLE BISECTORS OF A RECTANGLE TRICK

Suppose your friend would like to see how you can extend this trick to end up with a square. In that case, the trickster would start off with a rectangle instead of a general quadrilateral and once again draw the four angle bisectors, as shown in Figure 2.29. The result is then square *EFJG*.

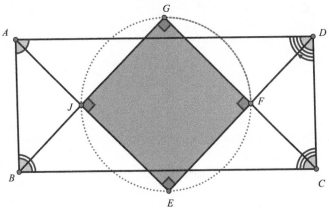

Figure 2.29

THE TRICK OF HOW TWO ANGLE BISECTORS GENERATE A RHOMBUS

This trick can be most effective with an audience of several people, each of whom has the capability of bisecting an angle. Have the members of the audience each draw a circle and inscribe in that circle a randomly drawn quadrilateral. Then have them extend the opposite sides of the quadrilateral to meet outside the circle, as shown in Figure 2.30, where we have the original simple quadrilateral *FBCE* inscribed in a circle. By extending the opposite sides and considering their points of intersection, *A* and *D*, we will have what is called a complete quadrilateral, *AFBCDE*. When we draw the bisectors of angles ∠*BAC* and ∠*BDF*, what results is truly unexpected. The points of intersection with the four sides of the original quadrilateral determine a rhombus, *GHJK*. Your audience will be truly amazed that everyone's diagram, regardless of the original shape of their quadrilateral, ends up with a rhombus.

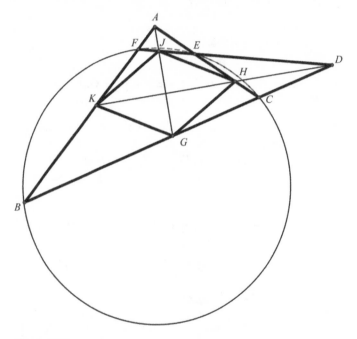

Figure 2.30

THE TRICK TO GET PARALLEL LINES

Surprise your friend with this trick, which will have them draw a line parallel to a side of a triangle without intentionally planning it. Have your friends draw any shape triangle and draw the median, that is, the line joining a vertex of the triangle to the midpoint of the opposite side, as we have done with line *AM* in Figure 2.31. Then ask them to merely draw *any* two lines that intersect each other on the median that they just drew. The points *D* and *E* at which these two lines intersect the two sides *AB* and *AC* of triangle *ABC* will determine

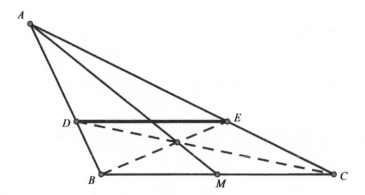

Figure 2.31

a line, *DE*, which amazingly is parallel to side *BC*. This trick can be useful for constructing a line parallel to a given line, where all you need is any triangle and its median to begin the construction.

THE UNEXPECTED PARALLEL LINES TRICK

Here is another trick that shows that parallel lines can evolve when they are least expected. Have your audience draw any triangle and locate the midpoints of the sides. Have them follow along as we now proceed. In Figure 2.32 we

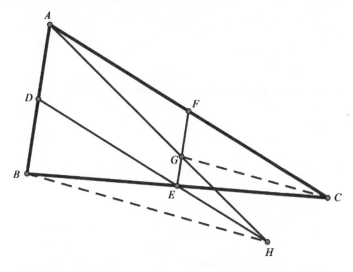

Figure 2.32

begin with triangle *ABC* and mark the midpoints of its sides at points *D, E,* and *F*. Have the audience select any point on *EF* and call it *G*. We then draw a line from *A* through *G* to meet *DE* at point *H*. Now, rather unexpectedly, the lines *GC* and *BH* end up being parallel. This once again shows the beauty of geometry, which often gets neglected when the subject is introduced.

THE SURPRISING PARALLEL LINES TRICK

Parallelograms often display unexpected properties, which can then be used in a trick to show the beauty of mathematics. Take, for example, parallelogram *ABCD*, shown in Figure 2.33, where any two parallel lines *AF* and *EC* are selected within the parallelogram, and where points *F* and *E* are located on sides *DC* and *AB*, respectively. From point *F* have your audience draw a line parallel to diagonal *AC* meeting side *AD* at point *P*. When they draw the line segment *PE*, they will be astonished to find that it is parallel to the other diagonal *BD*. Remember, points *F* and *E* could have been anywhere along the sides of the parallelogram as long as they generated the two parallel lines *AF* and *CE*. Quite a surprising result!

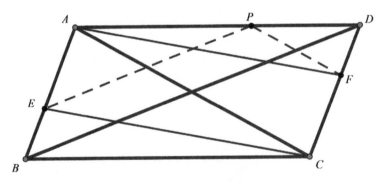

Figure 2.33

THE PARALLELOGRAM GENERATES CONCURRENCY TRICK

To do this trick, your friends need to start off with a parallelogram and extend one of its diagonals, preferably the shorter of the two. Then they will select any point on the extension of the diagonal and draw lines from that point to intersect the sides of the parallelogram, as shown in Figure 2.34. Here, we

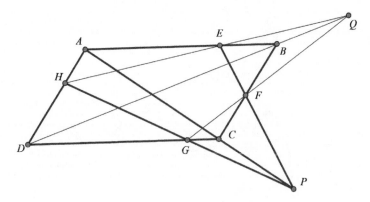

Figure 2.34

selected a point *P* on the extension of *AC*. From point *P* we then drew two lines intersecting the sides of parallelogram *ABCD* at points *E*, *F*, *G*, and *H*. Unexpectedly, we find that lines *HE*, *DB*, and *GF* are concurrent at point *Q*. This is a rather surprising trick, because of the randomness of the various points selected. Furthermore, notice that the concurrency point *Q* is on the extension of the diagonal *BD*—an extra little attraction.

CONSTRUCTING PERPENDICULAR LINES WITH ONLY A STRAIGHTEDGE

Challenging your audience to construct a line from a point outside a circle and perpendicular to a diameter of the circle would qualify you as a proper trickster. The audience would immediately wonder how this could be done without a pair of compasses. This is where the trickster will use a basic geometric theorem, attributed to the ancient Greek philosopher and mathematician Thales of Miletus (c. 624 BCE—c. 546 BCE): an angle inscribed in a semicircle is a right angle.

To do the construction, we will consider circle *O* and the point *P* outside the circle, as shown in Figure 2.35. Our task is to construct a perpendicular line from *P* to the circle's diameter *AB*.

We begin our task by joining the endpoints of *AB* and the point *P*. This will determine the intersection points with the circle, *C* and *D*. By drawing lines *AD* and *BC*, we create two lines that are altitudes, meeting at point *H*, of the triangle *PAB*, since *ACB* and *ADB* are both right angles according to Thales' theorem. Since we know that the altitudes of a triangle are concurrent, when we draw *PH* it will be perpendicular to *AB* since it will also be an altitude of triangle *PAB*. Thus, we will have constructed a perpendicular line from *P* to *AB*.

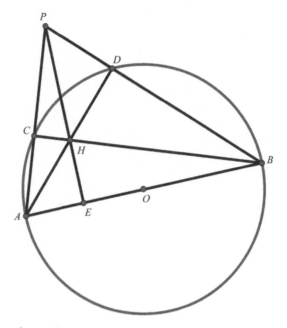

Figure 2.35

THE UNEXPECTED ISOSCELES TRIANGLE TRICK

This rather simple trick begins with one isosceles triangle and then unexpectedly results in another isosceles triangle. Begin by having your audience draw any isosceles triangle such as *ABC*, shown in Figure 2.36. Next, have them select any point *P* along the base *BC* of the triangle and erect a perpendicular line that would intersect the other two sides (extended) at points *D* and *E*. Unexpectedly, *ADE* will always turn out to be an isosceles triangle, with $AE = AD$. As simple as this trick is, it usually gets a "wow" reaction from the audience.

THE UNEXPECTED PERPENDICULAR-LINE TRICK

Begin this trick by having your audience draw a circle with two perpendicular lines, as shown in Figure 2.37, where *AC* is perpendicular to *BD*. Then have them draw the quadrilateral *ABCD*. An unexpected result occurs when a line is drawn from the midpoint, *F*, to the point of intersection of the two diagonals and extended to *AB*. We find that this line is always perpendicular to *AB*—quite unexpectedly!

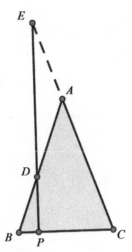

Figure 2.36

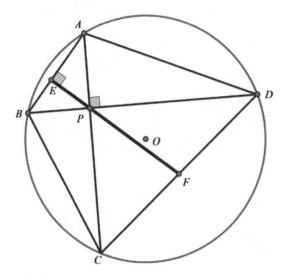

Figure 2.37

THE TRICK OF THE UNEXPECTED CONCURRENCY

Sit with a friend who has some mathematical tools at hand. Ask her to draw a circle, and then inscribe any randomly drawn quadrilateral in that circle. Your trick will be to guide her to come up with four lines that all go through one point—an unexpected concurrency.

First, have her locate the midpoints of the four sides of the quadrilateral. Then ask her to construct a line from each midpoint of the quadrilateral's sides and perpendicular to the opposite side. In Figure 2.38 we have drawn perpendiculars from the midpoints *E*, *F*, *G*, and *H* of the sides *AD*, *AB*, *BC*, and *CD*, respectively, to the opposite sides of the quadrilateral. Quite unexpectedly, these four lines are all concurrent at point *P.* This trick should clearly impress your friend, since there was no clue at all that such a concurrency would result.

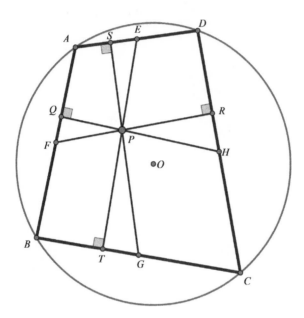

Figure 2.38

THE TRICK OF THREE COLLINEAR POINTS GENERATED BY A HEXAGON

It is best to involve several people working independently as you guide them through the trick. Each person should draw a circle and place six points on the circle, creating a hexagon with no opposite sides parallel. They should then extend the opposite sides until they meet. The three points of intersection in each of the audience's diagrams will all lie on the same line, or in other words, the three points are collinear. We show this in Figure 2.39, where the opposite sides *AF* and *DC* meet at point *N*; *EF* and *BC* meet at point *M*; and lastly, *ED* and *AB* meet at point *L*. Quite unexpectedly, we find that points *N*, *M*, and *L* are collinear; that is, they

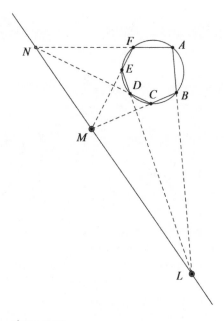

Figure 2.39

lie on the same straight line. This will be seen as quite a surprising result that the trickster has created. However, this amazing relationship was discovered by French mathematician Blaise Pascal (1623–1662), who published it at age 16.

THE TRICK OF THREE COLLINEAR POINTS GENERATED BY A HEXAGON

Now if you really want to impress your audience, you can take the previous trick one step further. Move the six points, which were the vertices of the hexagon along the circle, to different positions, but always remembering what the opposite sides of the original hexagon were.

Referring back to Figure 2.39, we note that the opposite sides of the hexagon *ABCDEF* are as follows: *AF* is opposite *CD*; *AF* is opposite *CD*; and *AF* is opposite *CD*. We will now place these points randomly on a circle, as shown in Figure 2.40. We notice that the pairs of the previously determined opposite sides intersect as follows:

AF and *CD* intersect at point *N*;
BC and *FE* intersect at point *M*;
AB and *ED* intersect at point *L*.

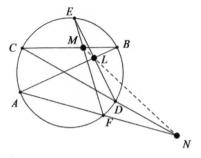

Figure 2.40

We will now take these intersections of the opposite sides as we did before and, if they exist (that is, "opposite sides" here should not be parallel), then they will be able to intersect. Once again, to our surprise (and awe), they result in three collinear points. We show this in the carefully constructed version of Figure 2.40, where points *N*, *L*, and *M* are, in fact, collinear. This will show the dynamic feature of geometry and can be demonstrated very nicely with dynamic geometry software programs such as Geometer's Sketchpad or GeoGebra.

THE TRICK OF THE UNEXPECTED COLLINEARITY

Once again, we have a trick that will require your audience to draw a perpendicular line. They can do this either with construction tools or a computer dynamic geometry program. What makes this trick so interesting is that your audience can easily draw a circle and inscribe any shape triangle in the circle. They then need to select any point on the circle, but not at one of the three vertices of the triangle. They are then asked to draw perpendicular lines from the selected point on the circle to each of the three sides of the triangle. This is shown in Figure 2.41, where *P* is any point on the circumscribed circle of triangle *ABC*. We then draw *PY* perpendicular to *AC* at *Y*, *PZ* perpendicular to *AB* at *Z*, and *PX* perpendicular to *BC* at *X*. The amazing thing about this trick is that regardless of the shape of the triangle and the placement of the point on the circle, the three points *X*, *Y*, and *Z* will always be collinear. This surprising result is today called the *Simson line*, incorrectly attributed to Scottish mathematician Robert Simson (1687–1768), but actually first discovered by the Scottish mathematician William Wallace (1768–1843) in 1799.

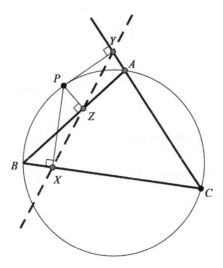

Figure 2.41

THE TRICK OF HAVING ONE CONCURRENCY GENERATE ANOTHER CONCURRENCY

To perform this trick your friends need to be able to draw a circle and have a straightedge available. Have them draw a circle and any three concurrent lines with endpoints on the circle, as shown in Figure 2.42, with the lines *JD*, *EL*, and *FG*, which meet at point *P*. Then tell your friends to construct a triangle whose sides contain points *D*, *E*, and *F*, and where each side intersects the circle at one other point, as shown in Figure 2.42 with the new intersection points *X*, *Y*, and *Z*. In this trick, when we draw the lines joining these last three points to the opposite vertices, surprisingly, the three new lines, *AY*, *BX*, and *CZ*, also will be concurrent. This is shown in Figure 2.43. A careful drawing by your friends should result in the same concurrency. This shows some unusual aspects of geometry, where one concurrency generates another concurrency.

THE TRICK OF THE SURPRISE EQUILATERAL TRIANGLE

This is a most unusual trick, but also rather difficult to do. It requires having your audience trisect an angle. It is well known that it is not possible to trisect a general angle with the normal construction tools: an unmarked straightedge

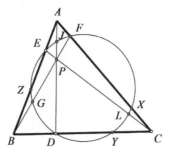

Figure 2.42

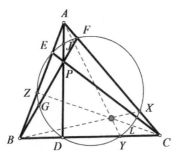

Figure 2.43

and a pair of compasses. Therefore, your audience will either have to estimate the trisection of an angle or perhaps use a protractor. In any case, you would begin by having your audience draw any scalene triangle, which would have no special properties. As you guide them through this trick, they will see that the angle trisectors of any triangle can determine an equilateral triangle.

In Figure 2.44 we have various triangles of different shapes, and in each case we have drawn the trisectors of its angles. We mark the intersections of adjacent trisectors as points *D*, *E*, and *F*. In each case, the triangle formed by these three intersection points *always* determines an equilateral triangle. The result from this trick should truly amaze the audience. Yet the proof, which can be done with nothing more than high school geometry, is a bit challenging and is referred to as Morley's theorem, attributed to American mathematician Frank Morley (1860–1937), who published it in 1900. Several proofs of this theorem can be found in *The Secrets of Triangles*, by A. S. Posamentier and I. Lehmann (Prometheus Books, 2012), pp. 351–55.

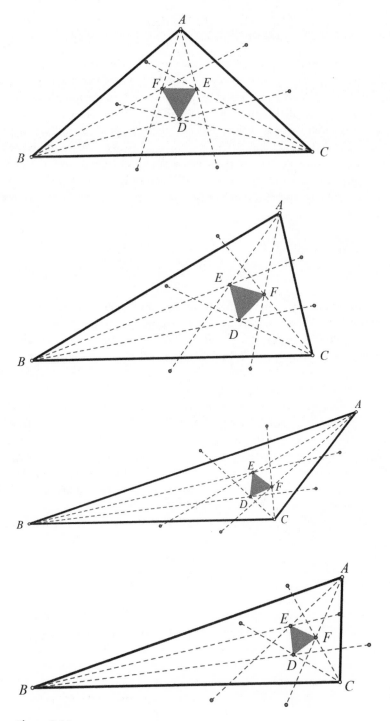

Figure 2.44

THE CUTTING TRICK

The simplicity of this trick tends to fool the audience, who expects a difficult challenge. Suppose you have a wooden cube whose edges are 3 inches in length. The challenge you present to your audience is to determine the least number of cuts needed with a saw to separate this cube into 9 cubes of edge 1 inch. You could suggest to the audience that they pile the cubes, as they cut them, in any way they choose. The trick here is that no matter how they are piled, six cuts will be necessary to partition the original cube into nine smaller cubes. Your audience will be surprised by how simple this is and by the fact that the center cube would require a cut on each of its six sides, therefore needing at least six saw cuts.

SPINNING AROUND A CYLINDER

This trick will surely cause your audience to think deeply about your talent as a trickster. Have them consider a cylinder whose circumference is 4 inches and whose height is 9 inches. Starting at a point at the top of the cylinder, wind a string helically around the cylinder so that after 10 times around, the other end of the string is at the bottom of the cylinder directly below the starting point. The challenge to the audience is to determine the length of the string.

The trick here is merely to roll out the cylinder on a plane and notice that the 10 revolutions represent a length of 40 inches and the length of the cylinder is 9 inches. This forms a right triangle, whose legs are 40 inches and 9 inches. Therefore, the hypotenuse, which is the length of the string, is 41 inches, which is obtained by applying the Pythagorean theorem to the right triangle. This may be a bit difficult for the audience to conceptualize. Yet, when given some time, they should be able to understand it.

A PAPER-FOLDING TRICK

It is difficult to construct a regular pentagon using an unmarked straightedge and a pair of compasses. However, a cute little trick for constructing a regular pentagon involves simply folding a strip of paper, say about 1"–2" wide, and making a knot. Then very carefully flatten the knot, as shown in Figure 2.45. Notice the resulting shape appears to be a regular pentagon—one with all angles congruent and all sides the same length. This can be better appreciated when you tear off the unneeded scraps on either end of the pentagon.

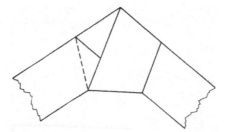

Figure 2.45

If you use relatively thin translucent paper and hold it up to a light, you should see a pentagon with its diagonals. These diagonals intersect in the golden ratio, as shown in Figure 2.46.

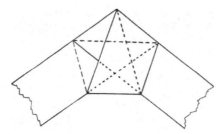

Figure 2.46

The golden ratio is perhaps the most famous ratio in geometry. It is typically seen on a rectangle, where the ratio of the width to the length is $\dfrac{W}{L} = \dfrac{L}{W+L}$ as shown in Figure 2.47. The numerical value of the golden ratio is then $\dfrac{1+\sqrt{5}}{2}$. To construct the golden ratio, we begin with square $ABCD$ and then extends the line AB from its midpoint the length of MC to create the rectangle $AEFD$, which is the golden rectangle.

Let's take a closer look at the regular pentagon in Figure 2.48, which evolved from our paperfolding trick. Point D divides AC into the golden ratio, since $\dfrac{DC}{AD} = \dfrac{AD}{AC}$.

Some readers might find it useful to see how the value of the golden ratio is embedded in this figure. To do this, begin with the isosceles triangle ABC, whose vertex angle has measure $36°$. Then consider the bisector BD, as shown in Figure 2.49.

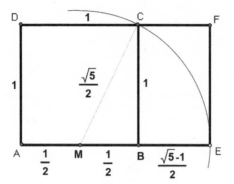

Figure 2.47

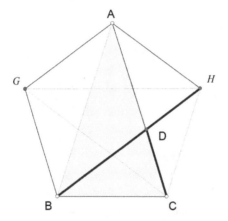

Figure 2.48

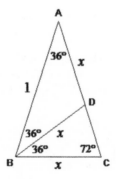

Figure 2.49

We find that $\angle DBC = 36°$. Therefore, $\triangle ABC \sim \triangle BCD$. Let $AD = x$ and $AB = 1$. However, since $\triangle ADB$ and $\triangle DBC$ are isosceles, $BC = BD = AD = x$. From the similarity above: $\dfrac{1-x}{x} = \dfrac{x}{1}$.

This gives us: $x^2 + x - 1 = 0$, and $x = \dfrac{\sqrt{5}-1}{2}$. (The negative root cannot be used for the length of AD.) Thus, $\dfrac{\sqrt{5}-1}{2} = \dfrac{1}{\phi}$. Point out to your audience that the golden ratio is the only number where the number and its reciprocal differ by 1. That is, $\phi - \dfrac{1}{\phi} = 1$, and for $\triangle ABC$ we have $\dfrac{\text{side}}{\text{base}} = \dfrac{1}{x} = \phi$, and, therefore, we call this a *golden triangle*. The motivated reader might want to delve more deeply into this famous ratio in *The Glorious Golden Ratio,* by A. S. Posamentier and I. Lehmann (Prometheus Books, 2012).

THE TRICK TO PROVE (?) THAT A RIGHT ANGLE IS EQUAL TO AN OBTUSE ANGLE

This geometric trick points out a few geometric properties that are often ignored. The audience will be tricked because they will not know about a rarely considered concept: a reflex angle. However, this trick will nevertheless leave them baffled. Follow along as we proceed to "prove" that a right angle can be equal to an obtuse angle (an angle that is greater than 90°).

We begin with a rectangle $ABCD$, where $FA = BA$, and where R is the midpoint of BC, and N is the midpoint of CF, as shown in Figure 2.50. We will now "prove" that *right* angle CDA is equal to *obtuse* angle FAD.

To set up the proof, we first draw RL perpendicular to CB, and draw MN perpendicular to CF. Then rays RL and MN intersect at point O. If they did not intersect, then RL and MN would be parallel, which would mean

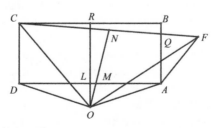

Figure 2.50

that *CB* would be parallel to, or coincide with, *CF*, which is impossible. To complete the diagram for our "proof," we draw the line segments *DO, CO, FO,* and *AO.*

We are now ready to embark on the "proof." Since *RO* is the perpendicular bisector of *CB* and *AD,* we know that *DO = AO*. Similarly, since *NO* is the perpendicular bisector of *CF,* we get *CO = FO.* Furthermore, since *FA = BA,* and *BA = CD,* we can conclude that *FA = CD*. This enables us to establish $\triangle CDO \cong \triangle FAO$ (SSS), so that $\angle ODC = \angle OAF$. We continue with *OD = OA,* which makes triangle *AOD* isosceles and the base angles *ODA* and *OAD* equal. Now, $\angle ODC - \angle ODA = \angle OAF - \angle OAD$ or $\angle CDA = \angle FAD$. This says that a right angle is equal to an obtuse angle. There must be some mistake! Let's see how the trick was performed.

Clearly, there is nothing wrong with this "proof." However, if you use a ruler and compasses to reconstruct the diagram, it will look like that shown in Figure 2.51.

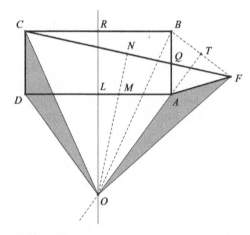

Figure 2.51

As you will see, the trick here rests with a reflex angle—one that is often not considered in the study of plane geometry. For rectangle *ABCD,* the perpendicular bisector of *AD* will also be the perpendicular bisector of *BC.* Therefore, *OC = OB, OC = OF*, and then *OB = OF*. Since both points *A* and *O* are equidistant from the endpoints of *BF,* the line *AO* must be the perpendicular bisector of *BF.* This is where the fault lies; we must consider the reflex angle of angle *BAO.* Although the triangles are congruent, our ability to subtract the specific angles no longer exists. Thus, the difficulty with this "proof" lies in its dependence upon an incorrectly drawn diagram, which is the basis for the trick.

THE TRICK TO PROVE THAT A TRIANGLE CAN HAVE *TWO* RIGHT ANGLES

This trick can truly upset an unsuspecting person. Make sure your audience is at ease before you embark on this upsetting trick. Begin with two intersecting circles of different, or the same, size, draw the diameters from one of their points of intersection (in this case *P*), and then connect the other ends of the diameters, here line *AB*, as shown in Figure 2.52.

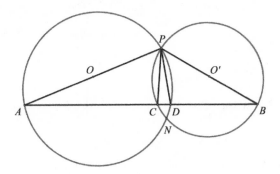

Figure 2.52

Here we have the endpoints of diameters *AP* and *BP* connected by line *AB*, which intersects circle *O* at point *D* and circle *O'* at point *C*. We find that ∠*ADP* is inscribed in semicircle *PNA*, and ∠*BCP* is inscribed in semicircle *PNB*. This makes them both right angles, since we know that any angle inscribed in a semicircle is a right angle. We then have tricked the audience by showing that triangle *CPD* has two right angles! This is impossible. Therefore, there must be a mistake somewhere in our work. Where were we tricked?

Omission of the concept of betweenness in Euclid's work could lead us to this dilemma. When we draw this figure correctly, we find that the angle *CPD* (in Figure 2.52) must equal 0, since the internal angles of a triangle cannot total more than 180°. That would make the triangle *CPD* nonexistent. Figure 2.53 shows the correct drawing of this situation.

In Figure 2.53 we can easily show that $\triangle POO' \cong \triangle NOO'$, and then $\angle POO' = \angle NOO'$. Because $\angle PON = \angle A + \angle ANO$, and $\angle ANO = \angle NOO'$ (alternate-interior angles), we have $\angle POO' = \angle A$, and then *AN* ∥ *OO'*. The same argument can be made for circle *O'* to get *BN* ∥ *OO'*. Since line segments *AN* and *BN* are each parallel to *OO'*, they must in fact be one line, *ANB*. This proves that the diagram in Figure 2.53 is correct and the diagram in Figure 2.52 is not. Therein lies the trick!

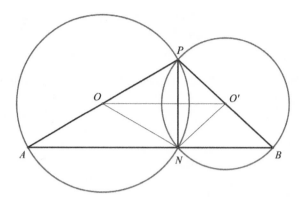

Figure 2.53

THE TRICK OF "PROVING" THAT ALL TRIANGLES ARE ISOSCELES

This trick will take some time and a patient audience who will probably be baffled by the results. However, we will clarify with a thorough discussion of where the trick led the audience astray.

We begin by drawing any scalene triangle (i.e., with no two sides of equal length) and then "prove" it is isosceles (i.e., with two sides of equal length). Consider the scalene triangle *ABC*, where we then draw the bisector of angle *C* and the perpendicular bisector of *AB*. From their point of intersection, *G*, we draw perpendiculars to *AC* and *CB*, connecting them at points *D* and *F*, respectively.

We now have four possibilities meeting the above description for various scalene triangles:

> In Figure 2.54, where *CG* and *GE* meet inside the triangle at point *G*.
> In Figure 2.55, where *CG* and *GE* meet on side *AB*. (Points *E* and *G* *coincide.*)
> In Figure 2.56, where *CG* and *GE* meet outside the triangle (in *G*), but the perpendiculars *GD* and *GF* intersect the segments *AC* and *CB* (at points *D* and *F*, respectively).
> In Figure 2.57, where *CG* and *GE* meet outside the triangle, but the perpendiculars *GD* and *GF* intersect the extensions of the sides *AC* and *CB* outside the triangle (in points *D* and *F*, respectively).

The trick can be done with any of the above figures. Follow along to see where you are being tricked. We begin with a *scalene* triangle, *ABC*. We will now "prove" that *AC* = *BC* (or that triangle *ABC* is isosceles).

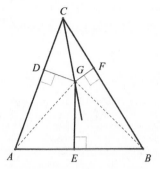

Figure 2.54

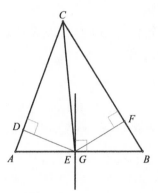

Figure 2.55

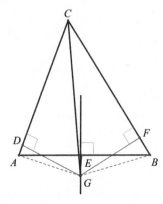

Figure 2.56

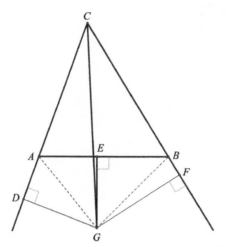

Figure 2.57

Because we have an angle bisector, we have $\angle ACG \cong \angle BCG$. We also have two right angles, such that $\angle CDG \cong \angle CFG$. This enables us to conclude that $\triangle CDG \cong \triangle CFG$ (SAA). Therefore, $DG = FG$, and $CD = CF$. Since a point on the perpendicular bisector (EG) of a line segment is equidistant from the endpoints of the line segment, $AG = BG$. Also, $\angle ADG$ and $\angle BFG$ are right angles. We then have $\triangle DAG \cong \triangle FBG$ (since they have respective hypotenuse and leg congruent). Therefore, $DA = FB$. It then follows that $AC = BC$ (by addition in Figures 2.54, 2.55, and 2.56; and by subtraction in Figure 2.57).

At this point you may feel thoroughly tricked. You may wonder where the error was committed, which permitted this trick to occur. You might challenge the correctness of the figures. Well, by rigorous construction you will find a subtle error in the figures. We will now divulge the mistake and see how it leads us to a better and more precise way of referring to geometric concepts.

First, we can show that the point G *must* be outside the triangle. Then, when perpendiculars meet the sides of the triangle, one will meet a side *between* the vertices, while the other will not.

We can "blame" this mistake on Euclid's neglect of the concept of betweenness. However, the beauty of this particular trick lies in the proof of this betweenness issue.

Begin by considering the circumcircle of triangle ABC (Figure 2.58). The bisector of angle ACB must contain the midpoint, M, of arc AB (since

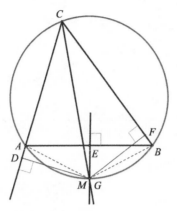

Figure 2.58

angles *ACM* and *BCM* are congruent inscribed angles). The perpendicular bisector of *AB* must bisect arc *AB* and therefore pass through *M*. Thus, the bisector of angle *ACB* and the perpendicular bisector of *AB* intersect on the circumscribed circle, which is *outside* the triangle at *M* (or *G*). This eliminates the possibilities we used in Figures 2.54 and 2.55.

Now consider the inscribed quadrilateral *ACBG*. Since the opposite angles of an inscribed (or cyclic) quadrilateral are supplementary, $\angle CAG + \angle CBG = 180°$. If angles *CAG* and *CBG* were right angles, then *CG* would be a diameter and triangle *ABC* would be isosceles. Therefore, since triangle *ABC* is scalene, angles *CAG* and *CBG* are not right angles. In this case, one must be acute and the other obtuse. Suppose angle *CBG* is acute and angle *CAG* is obtuse. Then in triangle *CBG* the altitude on *CB* must be *inside* the triangle, while in obtuse triangle *CAG*, the altitude on *AC* must be *outside* the triangle. The fact that one, and *only one*, of the perpendiculars intersects a side of the triangle *between* the vertices destroys the fallacious "proof." This demonstration hinges on the definition of betweenness, a concept not available to Euclid. Your audience should then be assured that not all triangles are isosceles, which, of course, is quite logical.

THE TRICK TO SHOW THAT 64 = 65

We now have a mathematics trick that was popularized by Charles Lutwidge Dodgson (1832–1898), who, under the pen name of Lewis Carroll, wrote *The Adventures of Alice in Wonderland*. In Figure 2.59 we notice that the square on the

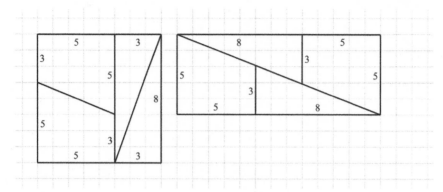

Figure 2.59

left side has an area of $8 \times 8 = 64$ and is partitioned into two congruent trapezoids and two congruent right triangles. Yet when these four parts are placed into a different configuration (as shown on the right side of this same figure), we get a rectangle whose area is $5 \times 13 = 65$. How can $64 = 65$? There must be a mistake somewhere. How can we be tricked in this way?

When we correctly construct the rectangle formed by the four parts of the square, we find there is an extra parallelogram in the figure—as shown, in exaggerated size, in Figure 2.60.

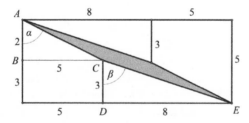

Figure 2.60

This parallelogram (shaded) results from the fact that the angles marked α and β are not equal. Yet, this is not easily noticeable at a glance in the original diagram. Perhaps the easiest way to show this is to refer to the familiar tangent function. In triangle *ABC*, $\tan \alpha = \dfrac{5}{2} = 2.5$, while $\tan \beta = \dfrac{8}{3} \approx 2.667$. In order for the line segment *ACE* to be a straight line—preventing a parallelogram from being formed—the angles α and β would have to be equal. With different tangent values this is not the case! Thus, the trick—easily overlooked—has been exposed.[2]

FORMING A SQUARE

Our natural thinking often leads us to comfortable solutions, and that is what this trick banks on. Then we hear the term "thinking out of the box" and we find that we must discard our normal thinking patterns. Here is a case in which we will be thinking geometrically in a somewhat different way. This may trick your audience at the end.

Suppose you have five congruent right triangles in which one leg is twice the length of the other. (See Figure 2.61.) The trick here is to form a square from these five right triangles, if you can make only one cut through one of the right triangles. At this point you let the audience think a bit before you provide the solution.

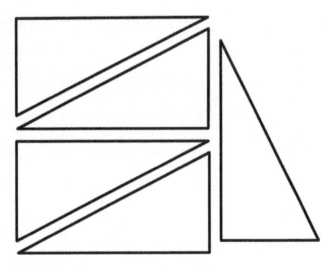

Figure 2.61

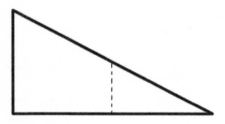

Figure 2.62

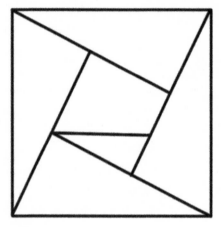

Figure 2.63

Trying to do this the traditional way will just lead the audience to move the pieces around in a fruitless effort. Yet, if we cut one of the right triangles along the perpendicular bisector of the longer leg as shown in Figure 2.62, and then place the pieces as shown in Figure 2.63, we will have formed a square, as we set out to do in the beginning.

Notice how the two acute angles of the cut-up triangle are complementary (their sum is 90°), so they form a right angle. Also, the two angles formed by cutting the hypotenuse are supplementary (their sum is 180°), which is also required for our final square. The solution should truly amaze your audience, yet allow them to see what it's like to "think out of the box."

WHERE IN THE WORLD ARE YOU?

Some entertainments in mathematics stretch the mind (gently, of course) in a very pleasant and satisfying way. Such examples leave us in amazement, which further generates our appreciation of mathematics. We present just such a situation now. This popular puzzle question, which has some very interesting extensions, requires some "out-of-the-box" thinking.

Ask your audience where on Earth you can be, so that, when you walk one mile **south**, then one mile **east**, and then one mile **north**, you end up at the starting point? (See Figure 2.64.)

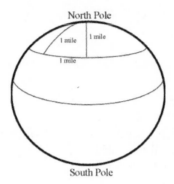

Figure 2.64

Most people will approach this problem using a trial-and-error method. After giving your audience some time to ponder, you might expose the trick to avoid frustration. We will now fortify the trickster with the correct answer. The answer is the North Pole. To test this answer, try starting from the North Pole and traveling south one mile. Then, traveling east one mile takes you along a latitudinal line, which remains equidistant from the North Pole, one mile south of it. Traveling one mile north gets you back to where you began, at the North Pole.

Upon discovering this answer, one usually feels a sense of satisfaction. Yet we can then ask: Are there other such starting points where we can take the three same-length "walks" and end up at the starting point? The answer, surprisingly enough, is *yes*.

One set of starting points is found by locating the latitudinal circle, which has a circumference of one mile and is nearest the South Pole. From this circle walk one mile north (along a great circle[3] route, naturally), and form another latitudinal circle. Any point along this second latitudinal circle will qualify. Try it now with your audience. (See Figure 2.65.)

Begin on this second latitudinal circle (the one farther north). Walk one mile south (takes you to the first latitudinal circle), then one mile east (takes you exactly once around the circle), and then one mile north (takes you back to the starting point).

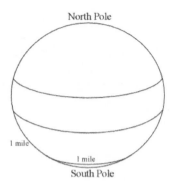

North Pole

1 mile

1 mile

South Pole

Figure 2.65

Suppose the second latitudinal circle, the one we would walk along, would have a circumference of ½ mile. We could still satisfy the given instructions this time by walking around the circle *twice*, and end up back at our original starting point. If the second latitudinal circle had a circumference of ¼ mile, then we would merely have to walk around this circle *four* times and then go north one mile to get back to the original starting point.

At this point we can make a generalization that will lead us to many more points that would satisfy the original stipulations (actually, an infinite number of points!). This set of points can be located by beginning with the latitudinal circle, located nearest the South Pole, which has a $\frac{1}{n}$-th-mile circumference. An n-mile walk east will take you back to the point on the latitudinal circle at which you began your walk. The rest is the same as before, that is walking one mile south and then later one mile north. Providing these various options should bring an "awakening"! This certainly will give the audience something to talk about.

THE TRICKY LINE THAT DIVIDES A TRIANGLE IN HALF

Here is a trick that will surely confound your audience. Have them draw any triangle and select a point on one side. The trick here is to draw a line from that point to another side so that the triangle is divided into two equal-area regions. The challenge here may overwhelm your audience; however, when the solution is presented, they will feel substantial relief, as well as awe!

The positioning of this line is somewhat unusual. Follow along as we construct the line and then justify why it divides the triangle into two equal areas. We will call our triangle *ABC*, as shown in Figure 2.66. We will then select

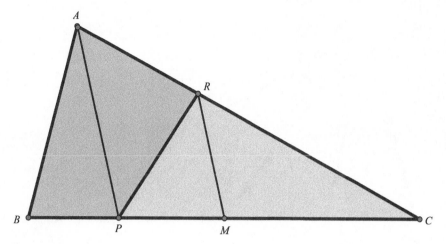

Figure 2.66

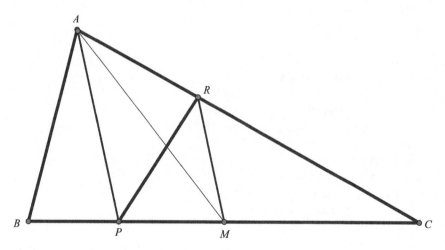

Figure 2.67

a random point P on side BC and then draw the line AP. The next step is to locate the midpoint of BC, which we will call point M. Through point M we will draw a line parallel to AP intersecting side AC at point R. This will now determine the sought-after line PR, which we will show divides the original triangle ABC into two equal-area regions.

To justify the claim that the line PR partitions the triangle ABC into two equal-area regions, we need to do the following: we draw a line AM, which we can easily show divides the triangle ABC into two equal-area triangles, ABM and ACM. (See Figure 2.67.) They are equal in area because they have a

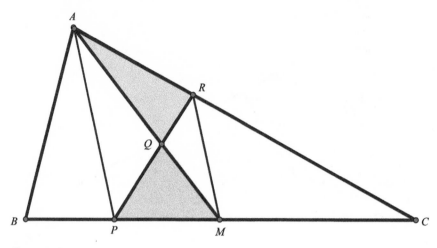

Figure 2.68

common altitude from A to BC, and they have equal basis since $BM = MC$.
Now comes the tricky part.

Consider the trapezoid $ARMP$, where we can show that triangle ARP is
equal in area to the triangle AMP. They share a common base AP and their
altitudes lie between parallel lines, which makes them equal in area. Refer now
to Figure 2.68. If we now remove triangle AQP from each of these two equal-
area triangles, we are left with two new equal-area triangles, ARQ and PMQ.
When we remove PQM from ABM and at the same time add its equal, ARQ,
alongside the remaining region of ABM, namely, the quadrilateral $AQPB$, we
then have established that the area of $ARPB$ is equal to the area of PRC. This
then has line PR partition ABC into two equal areas, as was our original goal.
This justifies the trickster's response.

GETTING TRICKED BY INFINITY

Here is where the trickster can have a lot of fun with his audience by simply
dazzling them with a slew of counterintuitive concepts. These all have to do
with our "understanding" of the concept of *infinity*, which is not a very easy
notion to grasp. However, presented in a lighthearted way, it could be entertaining and at the same time quite instructive. Follow along as we do a short
journey through everyday experiences that could involve thinking about infinity.

Perhaps one of the most curious measures of magnitude is the concept
that represents infinity, ∞. When we say that a set has an infinite number of

elements, to many people that means it has just a vast number of elements. Although, that is not incorrect, it is "short-selling" this concept. There are, in fact, orders of magnitude of infinity. For example, the set of all natural numbers {1, 2, 3, 4, 5, ...} is an infinite set. It is the same size as the set of all even numbers {2, 4, 6, 8, 10, ...}. This is already a concept that the trickster can bring to the stage to make people feel intellectually uncomfortable. Most people would find the equality of the set of all natural numbers and that of all even numbers hard to believe, since the set of natural numbers has all the elements of the set of even numbers as well as all the odd numbers. That logic would have one believe that the set of natural numbers is twice as large as the set of even numbers. But since the two sets are infinite, this is not true. Counterintuitive as it may seem, we can verify this claim of equality in the following way, which the trickster should get to after having the audience contemplate this "impossible inequality." We could argue that for every member of the set of natural numbers there will be an element in the set of even numbers to which it can match, thus making the two sets equal in size. That is, $1 \rightarrow 2$, $2 \rightarrow 4$, $3 \rightarrow 6$, $4 \rightarrow 8$, and so on. Of course, this works only when the sets are *infinitely* large. Clearly, if one takes the set of the first hundred natural numbers, that set is larger than the set of even numbers from 2 to 100. It is this concept of infinity that allows us to make this seemingly counterintuitive claim correctly. (One way to create a set larger than the set of natural numbers is to take the set of all subsets of natural numbers, which clearly would be a larger infinite set than the infinite set of natural numbers—but this is a bit beyond the scope of this book.)

At this point, the trickster should shift to geometry, since the concept of infinity also causes us discomfort in the geometric realm. Consider the following. We begin with a staircase, where each of the stairs can be of different sizes—although they could just as well be the same size, and it would not disturb our example. In Figure 2.69 we show a staircase where the sum of the vertical portions of each stair is a units, and the sum of the horizontal portions of each stair is b units. In other words, if we wanted to carpet the stairs from point P to point Q, we would require a length of carpeting $a + b$ units long.

In Figure 2.69 we see that the sum of the bold segments ("stairs"), found by summing all the horizontal and all the vertical segments, is $a + b$. If the number of stairs increases, the sum is still $a + b$, namely the sum of the height OP and the length OQ. Independent of the size of the steps, the sum remains constant, namely, $a + b$. Suppose we keep increasing the number of stairs, and as we do, the stairs naturally become smaller and smaller. At one point, the stairs will become so small that you will not even be able to see them as separate stairs; rather, the staircase will look like an oblique plane. Here is where our

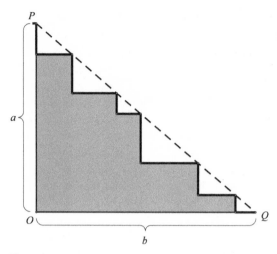

Figure 2.69

dilemma arises when we increase the stairs to a "limit" of infinity, so that the set of stairs appears to be a plane—or as seen from a side view, a straight line, which here is the hypotenuse of right triangle *POQ*. Following this line of reasoning, we would conclude that *PQ* has length $a + b$. Yet, we know from the

Pythagorean theorem that $PQ = \sqrt{a^2 + b^2}$ and *not* $a + b$. So, what's wrong? At this point, the trickster should allow the audience to ponder this dilemma. It is truly not an easy concept to comprehend.

While the set consisting of the stairs does indeed approach the straight line segment *PQ* as the number of stairs approaches infinity, it does *not*, therefore, follow that the *sum* of the bold (horizontal and vertical) lengths approaches the length of *PQ*, contrary to our intuition. There is no contradiction here, only a failure on the part of our intuition.

One way to explain this dilemma to the audience is to argue the following. As the stairs get smaller, they increase in number. In an extreme situation, we have 0-length dimensions (for the stairs) used an infinite number of times. This then leads to considering $0 \times$ "∞", which is meaningless! Please be warned that this is a very difficult concept for the audience to fully comprehend, and often takes much consideration before acceptance.

To further upset the audience, the trickster may wish to consider a similar argument in another geometric illustration. We begin with the semicircles shown in Figure 2.70, where the smaller semicircles extend from one end of the large semicircle's diameter to the other.

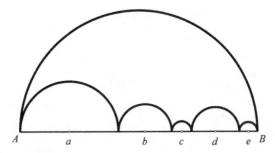

Figure 2.70

It is easy to show that the sum of the arc lengths of the smaller semicircles is equal to the arc length of the larger semicircle. That is, the sum of the smaller semicircles is $\dfrac{\pi a}{2} + \dfrac{\pi b}{2} + \dfrac{\pi c}{2} + \dfrac{\pi d}{2} + \dfrac{\pi e}{2} = \dfrac{\pi}{2}(a + b + c + d + e) = \dfrac{\pi}{2}(AB),$ which is also the arc length of the large semicircle. As a matter of fact, as we increase the number of smaller semicircles (where, of course, they get continuously smaller), the sum of the semicircle arcs remains the same. No matter how many semicircle arcs we fit in the space between points A and B, the sum of the arc lengths remains $\dfrac{\pi}{2} \times AB$. As the arcs get smaller and smaller, they will begin to disappear visually, and this arc length sum, $\dfrac{\pi}{2} \times AB$, "appears" to approach the length of segment AB, but, in fact, it does not. This would be absurd, since we know that $AB \neq \dfrac{\pi}{2} \times AB$, obviously!

Once again, we have a situation that could cause the trickster's audience some unease, since the set of arc lengths—consisting of the smaller semicircles—does indeed appear to approach the length of the straight-line segment AB. As we have shown above, it does *not* follow, however, that as the *sum* of the semicircles increases indefinitely, it approaches the *length* of the limit, which, in this case, is AB. This is one of the curiosities of the "number" infinity, ∞.

This "apparent limit sum" is absurd, since the shortest distance between points A and B is the length of segment AB, not the semicircle arc AB (which equals the sum of the smaller semicircles). This is an important concept to bear in mind so that future misinterpretations involving this curious concept of infinity, can be avoided. With this example, aside from upsetting the audience, they will begin to get some insight into this mysterious concept of infinity.

NOTES

1. These so-called *Müller-Lyer Illusions* were developed in 1889 by German psychiatrist Franz Müller-Lyer (1857–1916).
2. More such examples can be found in A. S. Posamentier and I. Lehmann, *The (Fabulous) Fibonacci Numbers*. Afterword by Herbert Hauptman, Nobel Laureate (New York: Prometheus Books, 2007), 140–43.
3. The great circle of a sphere is the largest circle that can be drawn on the surface of a sphere. It is formed by a plane intersecting the sphere that contains the center of the sphere.

3

Logical Thinking Tricks

As we have seen throughout the first two chapters of this book, mathematical tricks very often exhibit the beauty and power of mathematics. At the same time, they can be enriching and entertaining. Providing unexpected situations allows us to pose questions as tricks. There are also tricks that display a basis of mathematical thinking, often referred to as logical reasoning. In this chapter, we present a host of very simple questions, brainteasers, or puzzles, or what might be also called logic problems that will challenge the reader with tricky solution paths. To make the reading more entertaining, we separate this chapter into two parts. The first part presents the trick question, or conundrum. The second part provides the solutions, which for the most part expose the tricks, enabling us to truly appreciate logical reasoning and how mathematics benefits from this thinking. We intentionally distanced the solution from the question or proposed trick so that the reader does not intentionally glance at the solution before attempting to work it. The trickster also needs to appreciate these proposed tricks as a participant before presenting them to an audience.

TRICK 1: ANTICIPATING HEADS AND TAILS

This little, seemingly "impossibly difficult" trick will show you some clever reasoning along with how elementary algebraic knowledge can help you sort it. Let's begin by delving right into the question.

> *Your friend is seated at a table in a dark room. On the table there are 12 pennies, 5 of which are heads up and 7 are tails up. She knows where the coins are, so she can move or flip any of the coins, but because the room is dark, she will not know if*

*the coins that she is touching were originally heads up or tails up. You now ask her
to separate the coins into two piles of 5 and 7 coins and then flip all the coins in the
5-coin group. To everyone's amazement, when the lights are turned on there will be
an equal number of heads in each of the two piles.*

Your friend's first reaction is "you must be kidding! How can anyone do this
task without seeing which coins are heads up or tails up?" The solution will
surely enlighten the friend.

TRICK 2: THE HEAVY AND LIGHT COINS

Here is a trick problem to try to discern heavy coins from light coins. Pose
the problem as follows: arrange to get six coins of which three are heavy coins
and three are light coins. All look alike, and their weight cannot be visually
determined. Furthermore, the three heavy coins are all the same weight as are
the three light coins. Using only two weighings on a balance scale, how can
two of the three heavy coins be identified?

TRICK 3: THE COINS DILEMMA—WITH A
BALANCE SCALE

Yet another coin-identifying trick! Suppose you have nine coins that all look
the same and yet one coin is lighter than the other eight coins. Using a balance scale, how can you determine which is the lighter coin with only two
weighings?

TRICK 4: THE COINS DILEMMA—WITH A DIGITAL SCALE

Here is yet another good trick that can be quite entertaining. Consider having
10 stacks of 10 coins each. In 9 of the stacks, the coins all weigh 1 ounce. One
of the stacks has coins that weigh 2 ounces each. Challenge your audience to
identify the stack that contains the 2-ounce coins even though all the coins
look alike. The only tool they have available to do this identification is a digital
scale and only one weighing. Initial audience reaction is usually confusion or
bewilderment.

TRICK 5: THE SURPRISING NUMBER 22

At first, this trick will enchant the audience, and then (if properly presented) it will make them wonder why it works as it does. This is a wonderful opportunity to show off the usefulness of algebra, for it will be through algebra that their curiosity will be quenched.

Present the following to the audience.

> *Select any three-digit number with all digits different from one another. Write all possible two-digit numbers that can be formed from the three digits selected. Then divide their sum by the sum of the digits in the original three-digit number. You should all get the same answer, 22.*

This ought to elicit a big "WOW!" Let's consider the three-digit number 365. Take the sum of all the possible two-digit numbers that can be formed from these three digits. 36 + 35 + 63 + 53 + 65 + 56 = 308. Then we get the sum of the digits of the original number, which is 3 + 6 + 5 = 14. Then we divide 308 by 14 to get 22, which everyone should have gotten.

TRICK 6: THE COUNTERINTUITIVE TRICK

Some tricks present a challenge that seems impossible to meet. We are about to embark on one such trick now, one that has been often played in restaurants and bars using toothpicks. Suppose you are presented with a collection of toothpicks arranged as shown in Figure 3.1, where each of the two rows and two columns contains 11 toothpicks.

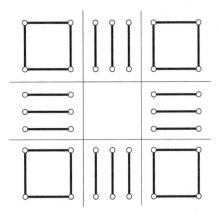

Figure 3.1

The challenge of this trick is that your audience is asked to remove one toothpick from each row and column and still have 11 toothpicks in each row and column. This seems to be impossible, since we are actually *removing* toothpicks, yet we are asked to keep the same number of toothpicks in each row and column, as before. Initially, the arrangement could look like that shown in Figure 3.2.

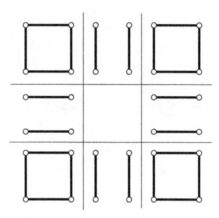

Figure 3.2

TRICK 7: THINKING OUT OF THE BOX

Here you give your friend a lattice of nine dots, as shown in Figure 3.3, and ask her to use four straight connected lines to touch all the nine dots, without lifting the pencil from the paper.

Figure 3.3

TRICK 8: PLACING THE COINS APPROPRIATELY

Give your friend ten coins and ask him to place them so that they form five straight lines, each of which contains four coins. The trick here is to realize that there is more than one solution. Allow your friend a bit of time before giving the solution.

TRICK 9: REVERSING THE BOWLING PINS TRIANGLE

Years ago, bowling pins were set up by pinboys and not by machine. Returning now to those old days, suppose the pinboy set up the pins in the wrong direction. Pose this predicament to your friend and ask him how the direction of the pins shown in Figure 3.4 can be reversed by moving only three pins?

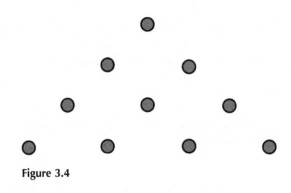

Figure 3.4

TRICK 10: CONNECTING THE 25 DOTS

In Figure 3.5 we show a 5 × 5 arrangement of dots. How can 12 of these dots be connected using straight lines to form a symmetric cross, with five dots inside it and eight dots outside? This trick, once again, requires thinking out of the box, which most audiences don't do.

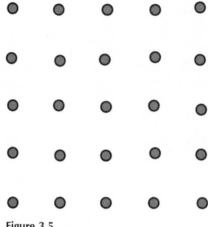

Figure 3.5

TRICK 11: THE TREE-PLANTING PROBLEM

This trick is easier said than done. The challenge here is to show how to plant 38 trees in 12 rows of trees with 7 in each row. A hint can be provided for your audience by telling them that they should avoid the traditional rectangular arrangement of trees.

TRICK 12: THE BROKEN CHAIN REPAIR PROBLEM

This trick can be done by providing your audience with four pieces of chain, each consisting of three links, as shown in Figure 3.6. On the other hand, you can do it in a virtual sense. The trick here is to show how these four pieces of chain could be made into a circular chain by opening and closing, *at most*, three links.

Figure 3.6

TRICK 13: THE SQUARE ORIENTATION

Here is a trick that will require a little thinking because of the wording in-volved. Suppose you are painting a square wall that measures 8 feet by 8 feet. When you are half finished, you tell your friend that the remaining unpainted half of the wall is still a square that measures 8 feet from top to bottom from the original top and 8 feet from side to side as well. How can this be done?

TRICK 14: THE SNAIL ESCAPING FROM THE WELL

Here the unsuspecting audience will give the wrong answer and be surprised by [or "at"] how tricky the problem is. A snail is at the bottom of a 10-foot well. During the day, he crawls up 3 feet. But at night he slides back 2 feet. So, clearly in one day he travels a distance of 1 foot in his quest to reach the top of the well. How many days will it take the snail to get out of the well?

TRICK 15: THE CLOCK-CHIMES

The beauty of this trick is that it pretty much stymies the audience. Very sim-ply said, the question goes as follows: At 5:00, the clock strikes 5 chimes in 5 seconds. How long will it take the same clock at the same rate to strike 10 chimes at 10:00? (Assume that the chime itself takes no time.) You can alert your audience that the answer is *not* 10 seconds, as most people immediately conclude.

TRICK 16: THE SOCKS IN THE DRAWER

Once again, we see how logical thinking can be perceived as a trick way of solving a problem. There are a number of such trick solutions that involve finding socks of a certain color in a drawer where you cannot see the colors. Consider, for example, the following tricks.

Variation 1:
In a drawer there are five blue socks, seven brown socks, and eight black socks. Without looking at the colors of the socks, how many socks would you have to take from the drawer to be certain of getting one pair of socks of the same color? A wide range of responses can be anticipated from your audience.

Variation 2:
Ask how many socks we have to draw from our above-mentioned drawer
to have in our hand at least one sock of each color.

Variation 3:
Suppose you have 12 identical black socks and 12 identical blue socks
in a drawer. How many socks would you have to take from the drawer,
without looking at them, in order to have a pair of blue socks to wear?

TRICK 17: DON'T WINE OVER THIS TRICK SOLUTION

The beauty of this trick rests in the elegant solution offered later; as unexpected as it is, it almost makes the problem trivial. However, our conventional
thinking patterns will likely cast a confusing haze over the problem. Don't allow your audience to despair. Encourage them to give it a genuine try, maybe
even struggle a bit. Here is the problem.

> *We have two one-gallon bottles. One contains a quart of red wine and the other, a quart*
> *of white wine. We take a tablespoonful of red wine and pour it into the white wine*
> *bottle. Then we take a tablespoon of this new mixture (white wine and red wine) and*
> *pour it into the bottle of red wine. Is there more red wine in the white wine bottle, or*
> *more white wine in the red wine bottle?*

TRICK 18: A TRICK SOLUTION

Problems are sometimes posed in social settings, because the solution can be
done in one step mentally. Yet the nature of such problems and the way most
people approach them is more a reflection of their earlier mathematics study.
In other words, the trickster will be able to impress the audience with an immediate solution, which most people seem to overlook.

The problem presented here can be easily solved by merely following the
"path" of the problem. This is something that most folks would do, as it is quite
natural. However, it does present a great opportunity to impress the audience
with a trick solution. You might wish to try the problem yourself (without
reading ahead) and see whether you are one of the majority who solve the
problem in a a somewhat inelegant fashion. The solution offered later will
probably enchant (as well as provide future guidance to) most readers.

> *The problem: A single-elimination (one loss and the team is eliminated) basketball*
> *tournament has 25 teams competing. How many games must be played until there*
> *is a single tournament champion?*

TRICK 19: LOGICAL THINKING

When a problem is posed that at first looks a bit daunting, and then a simple solution is presented—that seems to a be a trick solution. We often then wonder why we didn't think of that simple solution ourselves. Such problems that have a "gee-whiz" dramatic effect on us will likely help us with future (analogous) situations. Here is one example.

On a shelf in Danny's basement, there are three jars. One contains only nickels, one contains only dimes, and one contains a mixture of nickels and dimes. The three labels, "nickels," "dimes," and "mixed" fell off, and were all put back on the wrong jars (see Figure 3.7). Without looking, Danny can select one coin from one of the mislabeled jars and then correctly label all three jars. From which jar should Danny select the coin?

Nickels   **Dimes**   **Mixed**

Figure 3.7

TRICK 20: THE FLIGHT OF THE BUMBLEBEE

A trickster can present problems that lend themselves to very clever solutions. Often, after seeing the solution, we reflect over the "trick" solution with amazement. From such unusual approaches to a solution one learns problem-solving skills. Present the following to your audience.

The problem: *Two trains, serving the Chicago to New York route, a distance of 800 miles, start toward each other at the same time (along the same tracks). One train is traveling uniformly at 60 miles per hour and the other at 40 miles per hour. At the same time a bumblebee begins to fly at a speed of 80 miles per hour from the front of one of the trains toward the oncoming train. After touching the front of the second train, the bumblebee reverses direction and flies toward the first train (still at 80 miles per hour). The bumblebee continues this back-and-forth flying until the two trains collide, crushing the bumblebee. How many miles did the bumblebee fly before its demise?*

TRICK 21: WHERE IS THE MISSING MONEY?

Sometimes what appears to be true is not. We can see this with some simple business dealings, which can be posed as a problem leading to a trick solution. Suppose there are two vendors on the street selling belts. Each has 120 belts to sell. One sells them at two belts for $5, and the other vendor sells them at three belts for $5. The first vendor sold all of his belts and took in 60 × $5 = $300.

The second vendor also sold all of his belts and took in 40 × $5 = $200. Together they brought in $500.

The next week they decide to combine their 240 belts and sell them at the rate of five belts for $10—which they felt was the combination of their two previous prices (i.e., two belts for $5 and three belts for $5). When they sold all their belts, they took in 48 × $10 = 480. What happened to the $20 difference?

TRICK 22: LOGICAL BALANCING

This trick tests a person's ability to think counterintuitively. If a brick on one side of a balance scale balances evenly with the other side of the balance scale consisting of $\frac{3}{4}$ of a brick plus $\frac{3}{4}$ of a pound, then what is the weight of the brick?

TRICK 23: THE SURPRISING CUTS

This trick can be seen in a number of ways to a successful solution—logically and geometrically. Suppose you want to cut a pie into the maximum number of pieces with five straight-line cuts. What is the maximum number of pieces that can be made from the pie—they can be of different sizes—and where no piece gets moved during the cutting?

TRICK 24: THE UPSIDE-DOWN SQUARES

The trick here is to find the single number between 1 and 500 that is a square number and, when turned upside-down, is still a square number. You might mention to the audience that the only digits that can be read in both directions upside-down are: 0, 1, 6, 8, and 9. This way the original problem is more manageable.

TRICK 25: THE PLAYERS' DILEMMA

Here we offer a trick that you might want to try with two friends. It's a dilemma to figure out how much money each of three players had at the beginning of this game. The game is played by three players, and when a player loses, he must give each of the other two players double the amount of money that they have at that time. Each of the three players loses exactly once, and after these three games have been played, each player is holding $16. The question is, how much did each of the players have at the beginning?

TRICK 26: A TIRE-USE PROBLEM

The trick here is just to think logically. A car has traveled 100,000 miles and has used its four tires and a spare tire an equal amount, so five tires were used equally throughout the 100,000 miles. How many miles was each tire used?

TRICK 27: THE PILES OF COINS CHALLENGE

To show off your trick here, it might be best to actually set up the game in reality rather than just virtually as we've done so far. In this game there are three piles of coins on a table. One pile will have five coins, the second pile will have three coins, and the third pile will have one coin. A player can take any number of coins from any one pile during his or her turn. The person who takes the last coin on the table wins the game. The challenge here is to determine what moves the trickster should make to guarantee himself a win?

TRICK 28: THE 777 NUMBER TRICK

You might want to try this trick with a friend. Have your friend select any number between 500 and 1000; then have him add 777 to this number. If the sum exceeds 1000, then have him remove the thousands digit and add it to the units digit of the sum. Now have him subtract the two numbers—that is, this sum and the one he originally selected. Tell him that he must now have arrived at 222. How did this trick work?

Let's try one together now. Suppose your friend selects the number 600. He then adds to it 777 to get: $600 + 777 = 1377$. He now removes the 1 and adds it to the units digit to get 378. Subtracting these two numbers $600 - 378 = \mathbf{222}$.

TRICK 29: THE TRICKY NUMBER 9

We can construct many entertaining number tricks that involve the number 9. This is because 9 is 1 less than the base of our number system, 10. Many of these tricks rely on the fact that any number that is a multiple of 9 will have a digit sum that is a multiple of 9, which reduces, through continuous digit sums, to the number 9 itself.

For example, let's make up just such a number trick. Tell your friend to select any number and add his age to it. Then have him add to this result the last two digits of his telephone number and multiply this number by 18. He then sums the digits of his last number and continues to take the digit sum until he has a single digit. You impress him by telling him that this last number is 9. This should give your friend a bit of a surprise.

TRICK 30: BEATING THE CLOCK

Here is a trick that requires just a bit of logical thinking. A woman returning from work arrives regularly at her train station at 7 PM. One day the train comes in one hour earlier, arriving at the station at 6 PM. Knowing that her husband generally picks her up at 7 PM, she begins walking home, only to find her husband along the way, who drives her home the remaining distance, arriving home 10 minutes earlier than usual. We assume that her husband always meets her at exactly 7 PM when the train arrives. How long did the woman walk before her husband picked her up?

TRICK 31: THE EASILY CONFUSING SITUATION

A young fellow bought a football for $30. Needing money immediately, he sold it for $40. Sometime later, eager to play football, he bought the ball back for $50. After he finished playing, he found someone who wanted to buy it for $60. Did this young fellow make a profit or a loss in these transactions? How much was this profit or loss?

TRICK 32: A GUESS YOUR AGE TRICK

Here is a very simple trick that you can do with your friend. You will show him how you get his age by his following your instructions, which are as follows. Tell your friend to multiply his age by 3 and then add 6 to this product. Your friend tells you the result, and all you need to do is to subtract 2 from that number and you will have your friend's age.

TRICK 33: A PSYCHOLOGICAL TRICK

This trick has psychological aspects to it. You will simply ask a friend, who in this case is your target, to follow each number you orally give them by the next number in the sequence. For example, if you say 45, then he is to say 46. You then quickly give him 10 numbers such as 55, 13, 123, 6, 159, 225, 414, 1297, 68, and 4099. More than likely you will trick your friend into mistakenly saying 5000 as his last number, which clearly is incorrect.

TRICK 34: CREATING EQUAL AREAS

We now present a visual trick. How can four straight lines be drawn to partition the diagram consisting of three squares, shown in Figure 3.8, into four equal areas? Don't look at the solution until you have made a good effort to solve the problem. This will allow you to present it properly to your audience.

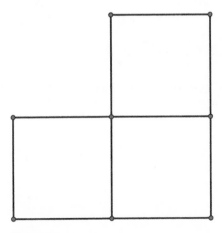

Figure 3.8

TRICK 35: TRYING TO CREATE EQUALITY

Max and Sam each have a bag of candies. They are now about to show their generosity. Max says to Sam: "If you (Sam) give me (Max) a certain number of candies from your bag, then I (Max) will have three times as many candies as you will have." If, however, Max gave Sam that same number of candies, then Sam would have half as many candies as Max would have. How many candies did each of the two boys have initially?

TRICK 36: A DECEPTIVE TRICK

This trick fools most people, unless they are suspicious about you, the trickster. If 5 men can plant 5 trees in 5 days, how many men will it take to plant 50 trees in 50 days?

TRICK 37: THOUGHTFUL REASONING

A trickster can assume that most people, when confronted with a problem, often resort to primitive ways of thinking. Sometimes these folks consciously think of analogous problems that they previously solved to see if these previous experiences can bring anything to the current problem. When primitive methods (those that can be called the "peasant's way") are used, a solution is unlikely, and if a correct solution does emerge, it will have taken considerably more time than an elegant solution (called the "poet's way," which results from thoughtful reasoning). A nice trick question that exhibits this thinking follows.

We are given a chessboard and 32 dominos, each of which is the exact size of two squares on the chessboard. The challenge is to ask the audience to show how 31 of these dominos can cover the chessboard, if a pair of opposite squares has been removed. (See Figure 3.9.)

Figure 3.9

TRICK 38: THOUGHTFUL REASONING—AGAIN

An analogous trick would be to ask your audience to determine a path through each of the rooms in the grid shown in Figure 3.10, where each room has a door on every interior wall. The path would start at the *entrance* and end at the *exit* at the diagonally opposite room. Emphasize that they must travel through each room before exiting.

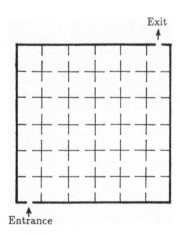

Figure 3.10

TRICK 39: AN ALGEBRAIC SURPRISE

One of the basic things we learned in secondary school algebra is to solve an equation. We also recall that when two variables are involved, a quadratic equation typically appears and requests a solution. In the following problem, this is very clearly signaled. If the audience does what was expected in the secondary school class, they will eventually end up with a quadratic equation. This, furthermore, is complicated by requiring the quadratic formula—and even having imaginary roots, which can easily distract and lead to a false solution. However, you will see, when you offer the solution shown in the solution section, that you have a trick that will truly amaze your audience. The problem is as follows: *If the sum of two numbers is 2, and the product of the same two numbers is 3, find the sum of the reciprocals of these two numbers.*

TRICK 40: THE TRICK OF THE HIDDEN SHORTCUT

With this problem you can demonstrate a technique that is too frequently overlooked, which will amuse your audience with its cleverness. Consider the following:

> *Two concentric circles are 10 units apart, as shown in Figure 3.11. What is the difference between the circumferences of the circles?*

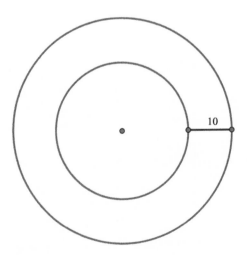

Figure 3.11

TRICK 41: OVERLOOKING THE OBVIOUS

A trickster presenting this problem's solution often elicits audience feelings of self-disappointment (for not having seen such an obvious solution). The problem seems to require using the Pythagorean theorem. We consider, in Figure 3.12, a point P on the circle with center O. Perpendicular lines are drawn from P to perpendicular segments AB and CD, meeting them at points F and E, respectively. Given that the diameter of the circle is 8, we are asked to find the length of EF.

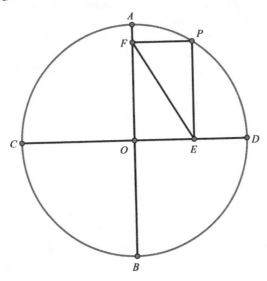

Figure 3.12

TRICK 42: WEIGHING THE WEIGHTS TRICK

This trick is a little bit longer than most and perhaps a bit more challenging. However, it does give the trickster the chance to show off. Suppose you have five weights of different sizes. The task is to arrange these five weights in order from heaviest to lightest, with seven weighings on a balance scale.

TRICK 43: THE TRICK TO WIN A NUMBER GAME

This game requires the trickster to know the allowable moves to win the game. The game begins by assigning the number 29 to each of the two players, one of which is the trickster. Each player is to select a square number to subtract from 29 and alternately subtract square numbers until the zero is reached. At no point should a subtraction result in a number less than zero. Whoever reaches zero first is the winner.

TRICK 44: THE PRIME STRATEGY

This trick presents a game that depends on knowing prime numbers (numbers whose only factors are 1 and the number itself) and has a strategy clearly marked out so that the trickster should be able to win each time. We begin by describing the game, where Jack and Charlie take turns removing toothpicks from a pile of 150. The rules of the game are each player takes at least 1 or a prime number of toothpicks at their turn and the person who takes the last toothpick wins the game. Let us assume Jack goes first. What is the strategy to win the game?

TRICK 45: PICKING TOOTHPICKS

Here the trickster plays a game against another person, which involves taking a specified number of toothpicks from a pile and trying to make the opponent pick the last toothpick, since he who picks the last toothpick loses the game. The rules of the game are that you can take only 1, 2, 3, 4, 5, 6, or 7 toothpicks at one time. The trickster should know the strategy to avoid being forced to take the last toothpick from a pile of 1000. That strategy is offered in the solution.

TRICK 46: THE LOGIC IN ALGEBRA TRICK

The trickster should propose the following algebraic problem to the audience. At first glance, the audience might be overwhelmed. But then a quick recollection of solving simultaneous equations in elementary algebra could direct them to a solution. However, the trick will make the trickster look powerful.

The following equations should be solved for each of the variables indicated:

$$x + y + z + u = 5$$
$$y + z + u + v = 1$$
$$z + u + v + x = 2$$
$$v + x + y + z = 4$$

TRICK 47: PLANTING IN ROWS

This problem can be solved by your audience very nicely with an algebraic trick. In this case, the trick will be nested in two equations. A farmer finds that he must plant 600 stalks of corn and would like to use a large number of irrigation ditches between rows. He realizes that if he takes five cornstalks from each of the rows he had already planned, he would be able to create six more rows than he originally had projected. The challenge here is to determine how many cornstalks were in his original plan. The audience may be confused about how to attack this problem. However, this is an excellent time to demonstrate the power of an algebraic solution. This trick should have a lasting effect on the audience because of its simplicity.

TRICK 48: REARRANGING THE FACE ON A CLOCK TO MAKE IT PRIME

Recall that a prime number is one whose only divisors are the number itself and 1. The trick here is to arrange the numbers on the face of a clock so that any six adjacent numbers are unmoved and the remaining six rearranged so that the sum of every pair of adjacent numbers is a prime number.

TRICK 49: TWO-WAY ADDITION TRICK

Figure 3.13 shows two numbers whose sums, added horizontally and vertically, all seem to be correct. In other words, $58 + 14 = 72$, and $15 + 48 = 63$, and the digits of the two sums each add up to 9. Furthermore, the three-digit numbers produce the horizontal sum $583 + 146 = 729$, and the vertical numbers provide the following correct sum, namely, $715 + 248 = 963$.

The trick is to find another set of numbers arranged so as to provide vertical and horizontal sums analogous to those demonstrated above.

5	8	**3**
1	4	**6**
7	**2**	**9**

Figure 3.13

TRICK 50: A PERFECT SUM

Here is a somewhat challenging problem the trickster can offer an audience. The divisors of the number 6 are 1, 2, 3, and 6. If we take the sum of the reciprocals of these four divisors, we get $\dfrac{1}{1} + \dfrac{1}{2} + \dfrac{1}{3} + \dfrac{1}{6} = 2$. The trick here is to find another set of reciprocals that likewise sum up to 2.

TRICK 51: THE SECRET AREA

This trick requires some geometric insight and will surely get a "gee-whiz" reaction from your audience. Figure 3.14 shows a square of sides length 9 and an isosceles right triangle whose equal sides are 12, so situated that the vertex of the right angle is at the center of the square. The question here is, what is the area of the shaded quadrilateral region?

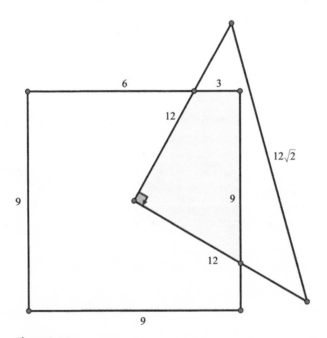

Figure 3.14

TRICK 52: REMOVING MATCHSTICKS TO MAKE
FEWER SQUARES

Moving matchsticks to make fewer squares is always a fun trick to show. Beginning with the 5 squares shown in Figure 3.15, the trick is to show how we can remove 2 matchsticks and be left with only 3 squares, and then show how it can be left with only 2 squares.

Figure 3.15

TRICK 53: MOVING MATCHSTICKS TO MAKE MORE SQUARES

Figure 3.16 shows 12 matchsticks placed to make 3 squares. The trick here is to show how 3 matchsticks can be moved into positions to create 5 squares.

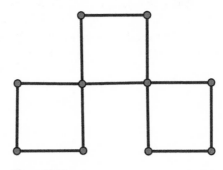

Figure 3.16

TRICK 54: REPRESENTING THE NUMBER ONE

The trickster can engage the audience with a challenge that seems difficult at the outset, but turns out to be rather simple. The challenge is to represent the number 1 using all of the numerals 2, 3, 4, 5, 6, 7, 8, and 9.

TRICK 55: AVERAGE SPEEDS

Suppose your friend drives from home to work at 30 mph due to heavy traffic. He then drives the same route back from work to home at 60 mph. What would you guess would be his average speed for the entire round trip? Most people would respond by taking the average of 30 and 60, as $\dfrac{30 + 60}{2} = 45$, but this is incorrect. This is where the trickster will clear up the problem by providing the right answer.

TRICK 56: CREATING A FAMOUS MAGIC SQUARE

Sometimes a trickster can not only perform a neat trick but also entertain the audience with the results. This is the case with magic squares. Perhaps the most famous magic square is one portrayed in a 1514 etching entitled "Melencolia I" by the famous German artist Albrecht Dürer (1471–1528) (Figure 3.17).

There are entire books written about magic squares[1] of all kinds. The one shown in Dürer's etching, however, stands out from among the rest. It has many properties beyond that required for a square matrix of numbers to be considered "magic." A magic square is a square matrix of numbers where the sum of the numbers in each of the columns, rows, and diagonals is the same. The Melencolia I etching, shown in Figure 3.17, exhibits a famous magic square in the upper right-hand corner, shown in enlarged form in figure 3.18.

Figure 3.17

Figure 3.18

Dürer signed most of his works with his initials, one over the other, along with the year in which the work was made. Here we find the signature near the lower right side of the picture, shown in Figure 3.19.

Figure 3.19

We notice that the etching was made in 1514. The audience should notice that the two center cells of the bottom row depict that year as well. We will look at this magic square more closely and show the trickster how easy it is to make a 4×4 magic square, and marveling over the many properties that this Dürer magic square has beyond the basics.

TRICK 57: CONSTRUCTING A BASIC ODD-ORDER MAGIC SQUARE

We have successfully mastered the trick of creating a simple even-order magic square and making it somewhat more enriching with lots of additional properties. We now show the trick in creating an odd-order magic square. The smallest odd-order magic square is 3× 3, as shown in Figure 3.20. The trick is to fill the 9 cells with the numbers from 1 to 9 so that each row, column, and diagonal has the same sum.

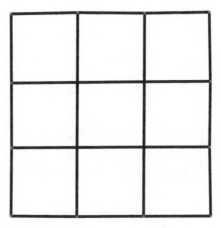

Figure 3.20

TRICK 58: CONSTRUCTING OTHER ODD-ORDER MAGIC SQUARES

Here is another opportunity for the trickster to extend her talents. Construct a 5×5 magic square, the next larger size after the previous one. The technique is the same, and the numbers now go from 1 to 25.

TRICK 59: CREATING THE RADIUS

Begin this trick by having your audience construct a semicircle and choose any point on the diameter, which in Figure 3.21 we call point *P*. At *P* have your audience construct two lines that form 60° angles with the diameter. Now ask your audience how the length of *AB* relates to the rest of the diagram. As trickster, you can eventually show them that the length of *AB* is equal to the radius of the semicircle. Remember, the amazing aspect here is that this is true for any point *P* on the diameter.

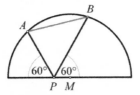

Figure 3.21

TRICK 60: BLACK AND RED CARDS IN A DECK

The trickster can test your logical thinking! A standard deck of 52 playing cards is randomly split into two piles with 26 cards each. How does the number of red cards in one pile compare to the number of black cards in the other pile?

SOLUTIONS

TRICK 1: ANTICIPATING HEADS AND TAILS

Here a most clever (yet incredibly simple) use of algebra will be the key to explaining the trick. Start with 12 coins, 5 with heads up and 7 with tails up. Have a friend separate the coins, without being able to look at them, into two piles of 5 and 7 coins each. Then she is to flip over the coins in the pile of 5 coins. Now both piles will have the same number of heads! That's all! She will think this is a magic trick. How did this happen?

Well, this is where algebra helps us understand what was actually done. When she separates the coins in the dark room, h heads will end up in the 7-coin pile. Then the 5-coin pile will have $5 - h$ heads. To get the number of tails in the 5-coin pile, we subtract the number of heads $(5 - h)$ from the total number of coins in the pile, 5, to get $5 - (5 - h) = h$ tails.

5-coin Pile	7-coin Pile
$5 - h$ heads	h heads
$5 - (5 - h) = h$ tails	
$= h$ tails	

When she flips all the coins in the smaller pile (the 5-coin pile), the $(5 - h)$ heads become tails and the h tails become heads. Now each pile contains h heads!

The piles after flipping the coins in the smaller pile

5-coin Pile	7-coin Pile
$5 - h$ tails	h heads
h heads	

This absolutely surprising result shows you how the simplest algebra can explain a very complicated trick.

TRICK 2: THE HEAVY AND LIGHT COINS

Here is what the trickster must do. Randomly selected three coins are placed on one side of the scale and the remaining three coins on the other side. Clearly, the side that has either two heavy coins or three heavy coins will certainly tip down. Two of the three coins on the down-tipped side are selected at random and are placed on either end of the balance scale. If the scale balances, then you have located the two heavy coins; that is, the two coins that were just weighed. If the scale does not balance, then the heavy coin will be on the down-tipped side and would partner with the third coin—omitted from this last weighing—as the two heavy coins. This trick can be done with actual coins or in a virtual situation.

Another coin-weighing problem should be easier now that the more difficult one above has given some guidance. This time you have two heavy coins of equal weight and two light coins of equal weight, and all four coins are visually indistinguishable. How can you identify the heavy coins and the light coins with only two weighings on a balance scale?

Using a similar trick reasoning as in the previous problem, we weigh any two of these four coins against each other on the balance scale. If the scale does not balance, then one heavy coin has been identified. All that needs to be done then is to weigh the two remaining coins to identify the second heavy coin. On the other hand, if the scale balances with the first two coins selected, then the two coins are either both heavy coins or light coins. All that needs to be done for the second weighing is to weigh one of these coins against the two remaining coins to determine whether the first pair of coins were both heavy or were both light. This will enable you to determine which two coins are the heavy coins. In either case, these coin-weighing problems can open up opportunities for other similar tricks.

TRICK 3: THE COINS DILEMMA—WITH A BALANCE SCALE

The trick here is to come up with a way to do the weighings, while being limited to two tries. After your audience becomes a bit frustrated, you can help them by suggesting that they take any three coins and weigh them against any other three coins. If the scale balances, then they will know that the lighter coin is among the remaining three coins that have not been weighed. They then take two of the three unweighed coins and place one on each end of the scale. If they balance, then the third coin—the one that has not yet been placed on the scale—is the lighter coin. If they don't balance, then the lighter coin will have revealed itself on the scale.

On the other hand, if the scale does not balance the two groups of three coins being weighed, then the lighter group of three will contain the lighter coin. In that case, use the procedure mentioned above to determine the lighter coin among the three in the lighter group.

TRICK 4: THE COINS DILEMMA—WITH A DIGITAL SCALE

This can be easily solved by numbering the 10 stacks of coins 1, 2, 3, ..., 9, 10, and representing each stack with the number of coins determined by their identifying number. So stack number 1 will have only one coin from its original 10 coins representing it; stack number 2 will have 2 coins from its original 10-coin stack representing it; stack number 3 will have 3 coins from its original 10-point stack, and so on, with stack number 10 represented by all 10 of its coins. These 10 newly defined stacks are placed on a digital scale. If all coins weighed the same, then the weight would be $1 + 2 + 3 + \ldots + 9 + 10 = 55$ ounces. However, we know that one of the stacks has coins that weigh 2 ounces each. Suppose that stack 5 was the stack with the 2-ounce coins. If that was the case, then our weighing would not have been 55 ounces but rather $55 + 5 = 60$ ounces. In other words, the difference between the actual weighing and 55 will determine which stack had the 2-ounce coins.

TRICK 5: THE SURPRISING NUMBER 22

Let's analyze this unusual result of everyone arriving at the number 22, regardless of which three-digit number they started with. We begin with a general representation of the selected number: $100x + 10y + z$. We now take the sum of all the two-digit numbers derived from the original three digits:

$$(10x + y) + (10y + x) + (10x + z) + (10z + x) + (10y + z) + (10z + y)$$
$$= 10(2x + 2y + 2z) + (2x + 2y + 2z)$$
$$= 11(2x + 2y + 2z)$$
$$= 22(x + y + z)$$

When this value, $22(x + y + z)$, is divided by the sum of the digits, $(x + y + z)$, the result is 22.

With this explanation, your audience ought to gain a genuine appreciation of how nicely algebra allows us to solve surprising tricks. Once again, we see how algebra explains a simple arithmetic phenomenon, and also exhibits its beauty.

TRICK 6: THE COUNTERINTUITIVE TRICK

Your audience then will ask, how can this possibly be done? After some contemplation your audience will see that if this can be done, they would have to count some toothpicks twice. Figure 3.22 shows that we have taken a toothpick from the center portion of each of the rows and the columns and placed these toothpicks in the corner position so that they could be counted more than once. This is the crux of the trick!

Thus, we have achieved our goal of having 11 toothpicks in each of the two rows and each of the two columns. This is a skill that merits attention, since it enables one to analyze real-life situations in a more critical fashion.

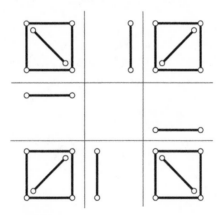

Figure 3.22

TRICK 7: THINKING OUT OF THE BOX

The typical attempts are to connect the dots along the sides and then notice that the four lines do not include the center dot. The trick here is to realize that you have to "think out of the box." In other words, your friend does not have to be restricted to stay on or within the square formed by the nine dots. Figure 3.23 shows one possible solution to this trick question. This trick can enlighten the audience to problem-solving thinking.

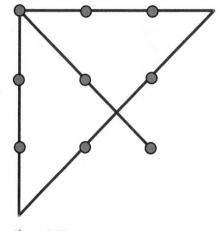

Figure 3.23

TRICK 8: PLACING THE COINS APPROPRIATELY

In Figure 3.24 we offer some possible solutions. In the diagrams each of the five lines contains four dots representing the coins. The trickster might want to encourage the audience to discover other options.

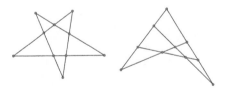

Figure 3.24

TRICK 9: REVERSING THE BOWLING PINS TRIANGLE

After multiple attempts to perform this task, the trickster might want to show how to move just three pins and reverse the direction of the triangle, which can be seen in Figure 3.25.

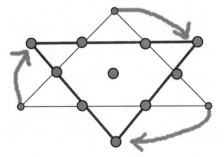

Figure 3.25

TRICK 10: CONNECTING THE 25 DOTS

The normal mindset is to attempt to draw this cross in the vertical and horizontal position. However, the trickster could suggest that the audience depart from this usual intuition. After your audience is thoroughly frustrated, you can present them with the correct solution, shown in Figure 3.26.

Tricks of this sort will open up a person's view of geometry beyond the typically anticipated moves.

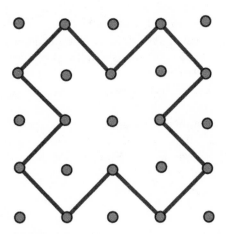

Figure 3.26

TRICK 11: THE TREE-PLANTING PROBLEM

The solution to this trick is best exhibited by the diagram shown in Figure 3.27, where we see 12 rows of trees, each having 7 trees—naturally, some trees are counted twice.

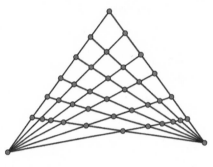

Figure 3.27

TRICK 12: THE BROKEN CHAIN REPAIR PROBLEM

Typically, the first solution attempt involves opening the end link of one chain, joining it to the second chain to form a six-link chain, then opening and closing a link in the third chain and joining it to the six-link chain to form a nine-link chain. By opening and closing a link in the fourth chain and joining it to the nine-link chain, a twelve-link chain is obtained, but it is *not* a circle. Thus, this typical attempt ends unsuccessfully. Most attempts usually involve other combinations of opening/closing one link of each of various chain pieces and trying to join them together to get the desired result, but these approaches will not work.

Let's look at this from another point of view. Instead of continually trying to open and close *one* link of each chain piece, a different point of view would involve opening *all* the links in one chain and using those links to connect the remaining three chain pieces together into the required circle chain. This trick quickly gives us the successful solution.

TRICK 13: THE SQUARE ORIENTATION

The trick here lies in the orientation of the square, which most people, from a psychological point of view, do not expect at first glance. Figure 3.28 shows the solution.

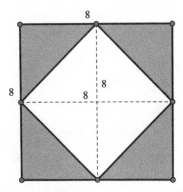

Figure 3.28

You will notice in this figure that the distance from top to bottom and side to side is still 8 feet, which is what the trick called for. Your audience will probably be quite surprised at how simple the solution was and yet eluded most of them.

TRICK 14: THE SNAIL ESCAPING FROM THE WELL

The typical response to this question is that the snail would exit after 10 days. However, the trick here is to realize that at the end of the seventh day, the snail will have traveled 7 feet. Then on the morning of the eighth day, the snail will crawl 3 feet and reach the top of the well. So the snail will not fall back that night, having reached the goal. In short, the snail will crawl out on the eighth day.

TRICK 15: THE CLOCK-CHIMES

Perhaps it is easiest to explain the correct answer by using a drawing to show exactly what is taking place. In the drawing in Figure 3.29, each circle represents a chime. Thus, the total time is 5 seconds, where the chimes themselves require no time, and there are four intervals between chimes. Therefore, each interval must take $\frac{5}{4}$ seconds.

Figure 3.29

Now let's examine the second case at 10 o'clock that will be *nine* intervals between the chimes, where each interval takes $\frac{5}{4}$ seconds. Therefore, the clock striking at 10:00 will take $9 \times \frac{5}{4} = 11\frac{1}{4}$ seconds. This trick question will keep folks from jumping to conclusions without thinking.

TRICK 16: THE SOCKS IN THE DRAWER

Variation 1:
To be assured of having one pair of the same color, we need to consider the worst-case scenario: on the first three tries, we have been unsuccessful, in that we have selected one of each color. Therefore, on the fourth attempt, the sock you pull out must match one of the previous three. Thus, we must take four socks from the drawer to be guaranteed a match of two socks of the same color.

Variation 2:
The trick reasoning here once again invokes the worst-case scenario: on the first eight tries we were able to get eight black socks, on the next seven tries we pulled seven brown socks. On the sixteenth try, there is only one other option left, namely, the blue socks. Therefore, we needed sixteen tries to be assured of getting one of each color.

Variation 3:
Once again, we consider the worst-case scenario where the first 12 selections are black socks. The next two selections would then have two blue socks. Thus, 14 socks selected would guarantee that you have obtained a pair of blue socks. The key to this trick is the question of being *guaranteed* to have selected two blue socks.

These problems will enhance the reasoning skills of your audience, which improves critical thinking.

TRICK 17: DON'T WINE OVER THIS TRICK SOLUTION

We can figure this problem out in any of the usual ways—often referred to as "mixture problems" in the high school context—or we can use some clever logical reasoning or one might say a trick solution. We begin as follows: With the first "transport" of wine there is only red wine on the tablespoon. On the second "transport" of wine, there is as much white wine on the spoon as there is red wine in the "white-wine bottle." This may require your audience to think a bit, but most should get it soon.

The simplest intelligible solution and the one that demonstrates a powerful strategy is that of considering *extremes*. We use this kind of reasoning in everyday life when we resort to the option "such and such would occur in a worst-case scenario, so we can ...."

Let us now employ this strategy for the above problem. To do this, we will consider the tablespoonful quantity to be a bit larger. Clearly the outcome of this problem is independent of the quantity transported, as long as it stays consistent throughout. So, we will use an *extremely* large quantity. We'll let this quantity actually be the *entire* one quart. That is, following the instructions given in the problem statement, we will take the entire amount (one quart of red wine) and pour it into the white-wine bottle. This mixture is now 50% white wine and 50% red wine. We then pour one quart of this mixture back into the red-wine bottle. The mixture is now the same in both bottles. Therefore, there is as much white wine in the red-wine bottle as there is red wine in the white-wine bottle! The trick lies in this solution procedure.

We can employ another trick procedure by considering another extreme case, where the spoon doing the wine transporting has a zero quantity. In this case, the conclusion follows immediately: there is as much red wine in the white-wine bottle as there is white wine in the red-wine bottle, that is, zero! Carefully presented, this solution can help people approach future mathematics problems and even enhance how they analyze their everyday decision making.

TRICK 18: A TRICK SOLUTION

Typically, the majority of problem solvers will begin to simulate the tournament by taking, for example, two groups of 12 teams playing each other during the first round, with one team drawing a bye (**12 games played**). After the first round, 12 teams will have been eliminated, leaving 12 teams and the one team that drew a bye remaining in the tournament. In the next round, of these 13 teams, 6 teams will play another 6 teams, leaving 6 winners and the one team that drew a bye (**6 games played**). In round 3, of the 7 teams remaining, 3 will play another 3, leaving 3 winners, plus the team that drew a bye (**3 games played**). In round 4, the 4 remaining teams will play each other (**2 games played**), leaving 2 remaining teams that will play each other for the championship (**1 game played**). They then count the number of games played, $12 + 6 + 3 + 2 + 1 = 24$ to get a champion team. This is a perfectly legitimate problem-solving technique, but clearly not the most elegant. Let us now consider the following procedure—the trick solution—to solve the original problem.

A much simpler way to solve this problem, one that most people do not naturally come up with as a first attempt, is to focus only on the losers and not on the winners, as was done above. Ask your audience the key question: "How many losers must there be in the tournament with 25 teams in order for there to be one winner?" The answer is simple: 24 losers. How many games must be played to get 24 losers? Naturally, 24. So there you have the answer, very simply done. Now most people will ask themselves, "Why didn't I think of that?" The answer is, it was contrary to the type of training and experience they have had. Looking at the losers gave us a handy way to grasp the solution.

There is an interesting alternative to the above solution that can be put as follows. Suppose that of the 25 teams, there is clearly one that is vastly superior to all the others. We could have each of the remaining 24 teams play this superior team and, of course, lose each game. Here again, you can see that only 24 games are required to get the champion, in this case, the predetermined superior team.

TRICK 19: LOGICAL THINKING

We could reason that the "symmetry" of the problem dictates that whatever can be said about the jar mislabeled "nickels" could just as well have been said about the jar mislabeled "dimes." Thus, if Danny chooses a coin from either of these jars, the results would be the same.

You, therefore, should concentrate on what happens if Danny chooses from the jar mislabeled "mixed." Suppose he selects a nickel from the "mixed" jar. Since this jar is mislabeled, it cannot be the mixed jar and must be, in reality, the nickel jar. Since the jar marked "dimes" cannot really be dimes, it must be the "mixed" jar. This leaves the third jar to be the dimes jar. You are probably thinking how simple the problem is—now that you have the solution. It does demonstrate a certain beauty of logical thinking. Your audience should be duly impressed!

TRICK 20: THE FLIGHT OF THE BUMBLEBEE

As the trickster, you might want to play off the natural inclination to want to find the individual distances that the bumblebee traveled. An immediate reaction is to set up an equation based on the famous (from high school mathematics) relationship: "rate times time equals distance." However, this back-and-forth path is rather difficult to determine; that is, it would require considerable calculation. Just the thought of having to do this can cause serious frustration. You might tell your audience that even if you could determine each part of the bumblebee's flight, the problem would still be very difficult to solve.

This is where the trickster can surely impress the audience. A much more elegant approach would be to look at the problem from a different point of view. We seek to find the *distance* the bumblebee traveled. If we knew the *time* the bumblebee traveled, we could determine the bumblebee's distance because we already know the *speed* of the bumblebee. Having two parts of the equation "rate × time = distance" will provide the third part. So, having the *time* and the *rate* will yield the distance traveled—albeit in various directions.

The time the bumblebee traveled can be easily calculated, since it traveled the entire time the two trains were traveling towards each other (until they collided). To determine the time, t, that the trains traveled, we need to set up an equation as follows: the distance traveled by the first train is $60t$, and the distance traveled by the second train is $40t$. The total distance the two trains traveled is 800 miles. Therefore, $60t + 40t = 800$, so $t = 8$ hours, which is also the time the bumblebee traveled. We now can find the distance the bumblebee traveled by again using the relationship rate × time = distance, which gives us $8 \cdot 80 = 640$ miles.

It is important to stress how to avoid falling into the trap of always trying to do what the problem calls for directly. At times a more circuitous method is much more efficient. Lots can be learned from this solution. You see, dramatic solutions often can be more useful than traditional solutions, since they provide an opportunity to "think out of the box."

TRICK 21: WHERE IS THE MISSING MONEY?

It appears that everything was perfectly logical. When they calculated their average rate of selling belts, they found the rates of sale 2 for \$5, or $\frac{2}{5}$, and 3 for \$5, or $\frac{3}{5}$ over the same number of belts. What they should have done is to divide the total number of belts by the total number of dollars: $\frac{240}{500} = \frac{12}{25}$, which would then have given the correct rate at which they should have sold belts in the combined form. They sold the 240 belts in the second week at the rate of $\dfrac{240}{480} = \dfrac{1}{2}$. So, you can see the difference in the selling price. Such confused calculations can make your audience wonder about their arithmetic abilities. But regardless of whether or not a calculator is used, the reasoning must be correct. Here the trickster has performed a lesson!

TRICK 22: LOGICAL BALANCING

Normally, the audience may see what one quarter of the brick would weigh and then find themselves frustrated. The trick here is to say that on the scale there would be a balance between the one whole brick and the combination of $\dfrac{3}{4}$ brick and pound weight, where the pound weight, which is $\dfrac{3}{4}$ of a pound, is there to make up for the $\dfrac{1}{4}$ brick missing from the $\dfrac{3}{4}$ brick. Therefore, the $\dfrac{1}{4}$ brick must weigh $\dfrac{3}{4}$ of a pound, and the brick must weigh $4 \times \dfrac{3}{4} = 3$ pounds.

TRICK 23: THE SURPRISING CUTS

There are several ways to approach this problem. The trickster can show that it could be easily solved geometrically as long as there are not any three cut lines concurrent. One such possibility is shown in Figure 3.30, where we notice that each cut intersects the four other cut lines.

Another possible solution to this trick is to set up a chart, as shown in Figure 3.31, where you keep track of the number of pieces added with each cut.

We notice that the number of cuts is equal to the number of pieces added with each cut. Taking this one step further than the problem requires, were we to make a sixth cut, we would be adding 6 additional pieces for a total of 22 pieces. The trick here is one of logical reasoning, as we saw in the geometric solution, and of looking for a pattern, as we saw in the chart.

TRICK 24: THE UPSIDE-DOWN SQUARES

The only square numbers (excluding the number 1) that can be formed using the digits 0, 1, 6, 8, and 9 are 9, 16, 81, 100, 169, and 196. Clearly the last two numbers of the list of squares are flips of one another, thus solving the problem.

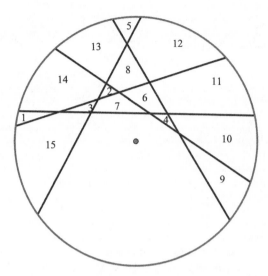

Figure 3.30

Number of cuts	Number of pieces	Number of added pieces
1	2	0
2	4	2
3	7	3
4	11	4
5	16	5

Figure 3.31

TRICK 25: THE PLAYERS' DILEMMA

The trick to solving this problem is to work backwards, starting with the amount of money each had at the end of the three games, namely, $16. Let's call our three players X, Y, and Z. The chart in Figure 3.32 traces the three games going backwards.

Amount of money each player has	X	Y	Z
At the end of three games	$16	$16	$16
Prior to the third game, which Z lost	$8	$8	$32
Prior to the second game, which Y lost	$4	$28	$16
Prior to the first game, which X lost	$26	$14	$8

Figure 3.32

The last row then shows what each player had before the three games began. To convince your audience that this is correct, have them play the game starting with the bottom row and working their way up to the top row.

TRICK 26: A TIRE-USE PROBLEM

Each of the four tires on a car traveling 10 miles will have been used equally. Therefore, the total number of tire miles for a car traveling 100,000 miles will be 400,000 miles of usage. When this total number of tire miles is divided by 5, we find that each tire will have been used for 80,000 miles.

Another trick solution would be to consider that one of the tires, the fifth tire, will replace each of the four tires for a distance of 20,000 miles. Therefore, each tire will have been used for 80,000 miles, since each tire had four positions to take at 20,000 miles per position.

TRICK 27: THE PILES OF COINS CHALLENGE

The trickster should take three coins from the pile holding five coins. This will leave the three piles as follows: Pile #1 has 2 coins; Pile #2 has 3 coins; Pile #3 has 1 coin. Any move that your opponent makes will allow you to leave either 2 piles of 2 coins each or 2 piles of 1 coin each. If your opponent takes all the coins from a pile of 2 coins, then you will win by taking all the coins from the remaining pile of 2 coins. Similarly, the same would hold true for the 2 piles of 1 coin each. On the other hand, if your opponent takes only 1 coin from a 2-coin pile, then you would take 1 coin from the other 2-coin pile. This leaves him 2 piles each having 1 coin, and since he must take coins from one pile, you would be left with the last coin to pick up. Once again, you, the trickster, would win the game.

Assume the trickster begins from the starting point of Pile #1 has 5 coins, Pile #2 has 3 coins, and Pile #3 has 1 coin. By taking any number of coins other than 3 from the 5-coin pile, your opponent would then get you into the losing situation of 3 piles having 3 coins, 2 coins, and 1 coin. You also could be left facing 2 piles of 3 coins, or 2 piles of 1 coin each, which also are losing positions for you.

TRICK 28: THE 777 NUMBER TRICK

Even as an experienced trickster, you may wonder why this works as it does. For every selected number between 500 and 1000 you will always get a 1 in the thousands place when this number is added to 777. Dropping the 1 and adding it to the units digit is tantamount to merely subtracting 999 from the number. That is, $-999 = -1000 + 1$.

If we now represent the selected number as n, here is what is being done:

$$n - (n + 777 - 999) = n - n - 777 + 999 = 222.$$

Remember, n represents the number that is randomly selected.

Suppose we would have used a number other than 777 as our "magic" number, say 591. Then we would have our friend end up with 408 every time, regardless of which number he chose between 500 and 1000. For a "magic" number of 733, the end result will always be 266. Remember the selected number cannot be less than 500, or else you might not get a sum in the thousands. At the same time the selected number should not be greater than 999, or you might get a 2 in the thousands place, which would ruin this scheme.

TRICK 29: THE TRICKY NUMBER 9

Rather than give a mere explanation, let's try this with the number 39 and add our age (37) to it to get 76. We then add the last two digits of our telephone number (31) to get 107 and then multiply this number by 18 to get 1926. The digit sum is $1 + 9 + 2 + 6 = 18$, whose digit sum is $1 + 8 = 9$. This works because when we multiplied by 18, we made sure that the final result would be a multiple of 9. This eventually always yields a digit sum of 9, since any number that is a multiple of 9 has a digit sum that is divisible by 9. And when we keep condensing the number by taking the digit sums we are always staying with a multiple of 9; therefore, the last number must be a 9.

TRICK 30: BEATING THE CLOCK

Let's take a look at the facts of the situation described. The couple arrived home 10 minutes earlier than usual. This implies that the husband drove to a point that was 5 minutes driving time from the station. Under normal circumstances, the husband would have been at that point at 6:55 PM. The wife started to walk at 6 PM; therefore, she must have been walking for 55 minutes before she was picked up by her husband.

TRICK 31: THE EASILY CONFUSING SITUATION

This trick is more confusing by its very nature and needs to be sorted out logically piece by piece. When this young fellow sold the football for the first time, he made a profit of $10. When he bought the football a second time, he made another profit of $10. Consequently, his profit for the entire set of transactions was $20. We should not be distracted from the difference in purchase price of the football on the two purchases, since that is irrelevant.

TRICK 32: A GUESS YOUR AGE TRICK

The best way to explain this trip is through algebra. If we assume that the person's age is represented by x, then to follow the instructions we would have to divide $3x + 6$ by 3, which is $x + 2$. Therefore, when you subtract 2 from this expression you are left with x, which is the required age sought.

TRICK 33: A PSYCHOLOGICAL TRICK

This trick is merely a psychological one in that you have tricked your friend into saying 5000 instead of 4100.

TRICK 34: CREATING EQUAL AREAS

As we know, a picture is worth a thousand words; therefore, we go directly to the diagram shown in Figure 3.33. The shaded region is formed by selecting the midpoints of the various sides of the three original squares. We then draw four lines as shown in Figure 3.33. Each of the three squares forming the shaded region is $\frac{1}{4}$ of the area of one of the three larger squares. Therefore, the remaining areas of each of the three original (larger) squares is $\frac{3}{4}$ of the original area of the square. Thus, we have four equal areas, each of which has an area $\frac{3}{4}$ of one of the original squares. The trick here lies in the unexpected construction.

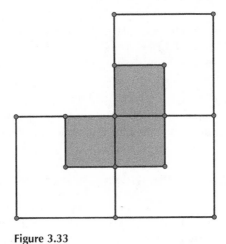

Figure 3.33

TRICK 35: TRYING TO CREATE EQUALITY

This trick will require us to use a bit of algebra. Begin by letting x equal the amount of candy that Max had in his bag, and let y equal the amount of candy that Sam had in his bag. We will let z equal the certain amount that is being transferred, which is being sought. Therefore, in the first transaction Max's additional candy would be equal to three times Sam's candy lot: $x + z = 3y - 3z$. The next option says that if Max gave Sam that amount of candy he would have half of what Sam has: $x - z = 2y + 2z$. Solving each of these equations for x, we get $x = 3y - 4z$, and $x = 2y + 3z$, respectively. From these two equations we get $y = 7z$. Substituting for y in either of the two equations gives us $x = 17z$, since z is an integer, the smallest of which is 1. Therefore, Max has 17 candies and Sam has 7 candies.

Let's see if this algebraic solution solves our little trick problem. Suppose Max has 17 candies and Sam 7 candies, and Sam gives Max 1 candy. Then Max has 18 candies compared to Sam's 6 candies, which means Max has 3 times as many candies as Sam has. If Max has 17 candies and Sam has 7 candies, then when Max gives Sam 1 candy he will only have 16 candies and Sam will have 8 candies. So Max will then have twice as many candies as Sam. Thus, we have solved the problem with algebra.

TRICK 36: A DECEPTIVE TRICK

The deceptive part of this trick is that the audience will assume the answer is 50 men, consistent with the original information. Since we know that 5 men can plant 5 trees in 5 days, then we can assume that 1 man can plant 1 tree in 5 days. If we multiply that by 10, then 1 man can plant 10 trees in 50 days. When we multiply that by 5, we find that 5 men can plant 50 trees in 50 days, which answers our original question. It is a rather tricky result and not easily anticipated, making this a true learning experience.

TRICK 37: THOUGHTFUL REASONING

As soon as the question is posed, most people try various arrangements of square covering. This may be done with actual tiles or with a graph grid drawn on paper and then shading adjacent squares two at a time. Before long, frustration begins to set in, since this approach cannot succeed. Here the issue is to go back to the original question. A careful reading of the question reveals that it does not say to do this tile covering; it asks *if it can* be done. Yet, because of the way we have been trained, the question is often misread as "do it."

A bit of clever insight helps. Ask yourself the question: "When a domino tile is placed on the chessboard, what kind of squares are covered?" A black square and a white square must be covered by each domino placed on the chessboard. Is there an equal number of black and white squares on the truncated chessboard? No! There are two fewer black squares than white squares. Therefore, it is impossible to cover the truncated chessboard with the 31 domino tiles, since there must be an equal number of black and white squares. The question has then been answered.

TRICK 38: THOUGHTFUL REASONING—AGAIN

Once again, the trick here is that such a path does not exist, where one enters at the entrance, travels through each room, and exits as indicated. The reasoning is very similar to the one on the chessboard previously shown, since the opposite corners, the entrance and exit rooms, have something in common.

Consider a map of our 36 rooms drawn so that adjacent rooms are of different colors. Figure 3.34 uses black and white to show that adjacent rooms are of different colors.

Figure 3.34

When walking from room to room, one has to travel from a white room to a black room. We continue in this way, never traveling through the same color room twice, consecutively. There is an equal number of black rooms and white rooms. The entrance is in a white room, so the exit would have to be in a black room. Since this is not the case, such a walk is impossible. Comparing this trick to the previous one should give the trickster ideas for further conundrums of this sort.

TRICK 39: AN ALGEBRAIC SURPRISE

The first step that most people would take—based on previous experience—is to set up the equations that follow from the problem statement: $x + y = 2$ and $xy = 3$. A typical next step would be to solve for y in the first equation to get $y = 2 - x$, and then to substitute that into the second equation to get

$x(2 - x) = 3$, which then can be written as $x^2 - 2x + 3 = 0$. Since the trinomial is not factorable, the quadratic formula $\left(x = \dfrac{-b \pm \sqrt{b^2 - 4ac}}{2a} \right)$ will be required, and the resulting roots will be imaginary numbers, $1 \pm i\sqrt{2}$. Once these numbers have been obtained, you would need to find the reciprocals and add them: $\dfrac{1}{1 + i\sqrt{2}} + \dfrac{1}{1 - i\sqrt{2}} = \dfrac{\left(1 + i\sqrt{2}\right) + \left(1 - i\sqrt{2}\right)}{\left(1 + i\sqrt{2}\right)\left(1 - i\sqrt{2}\right)} = \dfrac{2}{3}$. Having now seen the traditional way to solve this problem, as trickster you can now show a much more elegant solution—one that will amaze your audience.

Now for the trick solution. Remember, we were not asked to find the values of x and y. Rather, we were asked to find the sum of the reciprocals of these, namely, $\dfrac{1}{x}$ and $\dfrac{1}{y}$. So, it turns out that there is no need to find x and y. We know that the sum of these reciprocals is $\dfrac{1}{x} + \dfrac{1}{y} = \dfrac{x + y}{xy}$. Therefore, we have the answer immediately, since we know that $x + y = 2$ and $xy = 3$. We can substitute that into the equation and get the fraction $\dfrac{2}{3}$. And so we have gotten the answer to the original question immediately, avoiding a lot of messy algebra.

TRICK 40: THE TRICK OF THE HIDDEN SHORTCUT

The traditional straightforward method for solving this problem is to find the diameter and the circumference of the two circles. Subtracting the two circumferences will then give you the required answer. Since the diameters are not given, the problem becomes somewhat more complicated than usual. Therefore, we will let d represent the diameter of the smaller circle. $d + 20$ then is the diameter of the larger circle, since we have the diameter d of the smaller circle plus twice 10, the distance between the circles. The circumferences of the two circles will then be πd and $\pi(d + 20)$, respectively. The required answer is then obtained very easily by taking the difference of the circumferences, which is $\pi(d + 20) - \pi d = 20\pi$. (See Figure 3.35.)

Now the trickster can exhibit a talent with a more elegant and vastly more dramatic procedure, which involves using an extreme case. Since we were not told how large the circles are, we will let the smaller of the two circles become even smaller until it reaches an "extreme smallness," or a point. All we were given was the distance between the two circles, which we will preserve.

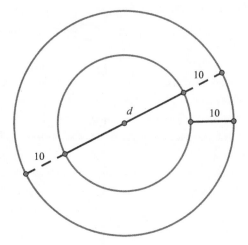

Figure 3.35

Therefore, we can consider the two circles to take on any convenient sizes, as long as we keep the distance between their circumferences to be 10 units. Since the smaller circle was reduced to a point, it coincides with the center of the larger circle. The distance between the two circles now becomes simply the radius of the larger circle. The difference between the lengths of the circumferences of the two circles, which was our original question, is now merely the circumference of the larger circle, or 20π. So once again, you will have impressed your audience with a trick solution. This example might serve them well in the future when solving a problem that can be reduced to a trivial one by considering an extreme situation.

TRICK 41: OVERLOOKING THE OBVIOUS

Since we see a right triangle in Figure 3.36, we might be tempted to use the Pythagorean theorem to solve the problem. However, the trickster, after having given the audience some time to mull over the problem, might point out that point *P* could be anywhere on the circle and further frustrate the audience as they seek a solution.

At this point the trickster might want to perform a quick trick that gives a trivial solution. Stepping back from the problem and looking at it in a fresh way will reveal that the quadrilateral *PFOE* is a rectangle, since the perpendiculars determine three of its right angles. We recall that the diagonals

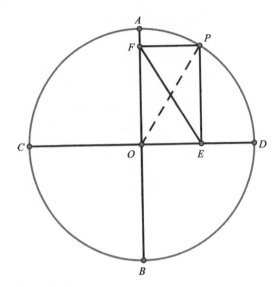

Figure 3.36

of a rectangle are equal in length; therefore, *FE* must equal *PO*, which is the radius of the circle and is half the length of the diameter, or 4.

The trickster can once again show talent by having the audience look at the problem in another fashion. Take the location of *P* at a more convenient point, say at point *A*. In this case, *FE* would coincide with *AO*, which is the radius of the circle.

TRICK 42: WEIGHING THE WEIGHTS TRICK

In order to facilitate an explanation of the trick's solution we will call the five weights A, B, C, D, and E. We begin by weighing A against B, and C against D. We then weigh the two heavier ones, A and C. At this point we will have had three of our seven allowed weighings. We now weigh E against C, and we assume that E is heavier than C. We now approach our fifth weighing, which will compare the weights of E and A. If A is heavier than E, we then weigh B against C, which is then followed by a weighing of B and E, or B and D depending upon whether B is heavier or lighter than C. If E is heavier than A, then we weigh B against C and, if necessary, weigh B against D. This should give you the order of the five weights in seven or less weighings. In like manner, we could determine the case where E is lighter than C.

TRICK 43: THE TRICK TO WIN A NUMBER GAME

The trickster should know how to win this game by being the first player and subtracting the square number 9. He then must follow the opponent's moves in the following way: if the opponent's next move is subtracting the 1, then the trickster would follow by subtracting 9. If the opponent subtracts 4, then the trickster follows by subtracting 16. If the opponent subtracts 9, then the trickster also subtracts 9. If the opponent subtracts 16, then the trickster subtracts 4.

Following this pattern, the trickster will always win, using his talent of knowing the trick!

TRICK 44: THE PRIME STRATEGY

Jack would win the game by taking 2 toothpicks from the pile of 150, which leaves 148. At each future turn, Jack would need to always leave a number of toothpicks in the pile that is a multiple of 4. Charlie can never take a multiple of 4, since it is not a prime number, which means that he cannot leave a multiple of 4 on the pile of toothpicks. At each turn, Jack chooses 1, 2, or 3 toothpicks so that he always leaves a pile that is a multiple of 4, from which Charlie is to select. Eventually, he will reach 0 and win the game.

TRICK 45: PICKING TOOTHPICKS

Naturally, the trickster must know how to win this game. To do so, you must figure out how he can leave his opponent one toothpick on the pile. The strategy should be to leave $8k + 1$ toothpicks on the pile after each of the trickster's turns. If the opponent takes x toothpicks on a turn, then the trickster can take $8 - x$ toothpicks on the next turn. The trickster's first turn should take 7 toothpicks, which leaves the opponent with $1000 - 7 = (8)(125) - 7 = (8)(124) + 1$ toothpicks. Continuing with the strategy, the trickster will eventually leave one toothpick to be taken by the opponent, who then will lose the game.

TRICK 46: THE LOGIC IN ALGEBRA TRICK

The trick to solving the set of equations is to add them and then divide by 4:

$$x + y + z + u = 5$$
$$y + z + u + v = 1$$
$$z + u + v + x = 2$$
$$v + x + y + z = 4$$
$$\underline{x + y + v + u = 0}$$
$$4x + 4y + 4z + 4u + 4v = 12$$

When we divide by 4, we get:

$$x + y + z + u + v = 3.$$

All we need to do now is to subtract each of the above equations from this last equation in order to get each of the variables. For example, when we subtract the first equation from this obtained equation, we get $(x + y + z + u + v = 3) - (x + y + z + u = 5)$ we then get $v = -2$. The remaining subtraction yields the following: $x = 2, y = 1, u = -1,$ and $z = 3$.

Although the problem seemed overwhelming at the start, the trickster has shown how it can be done in a flash!

TRICK 47: PLANTING IN ROWS

We begin by letting x equal the number of cornstalks in each row, and y the number of rows in the original proposal. Therefore, the product $xy = 600$ represents the original schematic. By removing five cornstalks from each row, which results in six additional rows, we get the equation $(x - 5)(y + 6) = 600$. From elementary algebra, we know that we must solve these two equations simultaneously, so from the second equation we get:

$xy - 5y + 6x - 30 = 600$. Then, substituting the first equation into the second equation, we get $600 - 5y + 6x - 30 = 600$, which gives us $-5y + 6x - 30 = 0$. From the first equation we know that $y = \dfrac{600}{x}$, which we then substitute into the previous equation to get $-5\left(\dfrac{600}{x}\right) + 6x = 30$, which can be simplified as $x^2 - 5x - 500 = (x - 25)(x + 20) = 0$. Therefore, $x = 25$, which represents the number of cornstalks in each row of the original schematic. The trick here is simply for the audience to realize that algebra can play a very important role in solving a problem.

TRICK 48: REARRANGING THE FACE ON A CLOCK TO MAKE IT PRIME

One possibility is to leave the numbers 11, 12, 1, 2, 3, and 4 intact and to arrange the numbers 7, 10, 9, 8, 5, and 6, as shown in Figure 3.37. Notice that each pair of numbers adds to a prime number, such as $7 + 10 = 17$, $9 + 10 = 19, 8 + 5 = 13$, and so on.

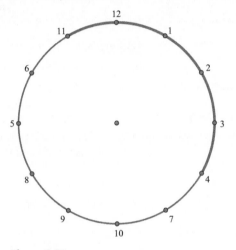

Figure 3.37

Another possibility for arranging the numbers is to keep the same six numbers, 11, 12, 1, 2, 3, 4, in place and rearrange the remaining numbers as follows: 9, 10, 7, 6, 5, 8, as shown in Figure 3.38. Once again, the sums of pairs of partner numbers will always be prime numbers.

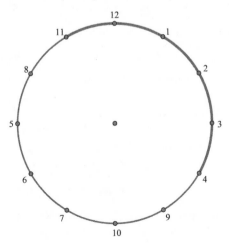

Figure 3.38

TRICK 49: TWO-WAY ADDITION TRICK

As a trickster you should know there is *only one other* pair of numbers that will make this vertical and horizontal addition work. It can be seen in Figure 3.39.

4	8	2
1	5	7
6	3	9

Figure 3.39

Here we can see the following additions. First, vertically the two-digit numbers yield $48 + 15 = 63$. And looking at this horizontally, the two-digit numbers provide the sum $14 + 58 = 72$. Once again, taking the sum of the three-digit numbers vertically we have $482 + 157 = 639$, and the horizontal sum is $614 + 358 = 972$. There are no other numbers that would satisfy this condition.

TRICK 50: A PERFECT SUM

Rather than just give you the correct answer, we show the beauty of this relationship by developing the answer algebraically. We begin by letting 1, a, b, c, …n represent the divisors of n in increasing order. We then make the supposition that $\dfrac{1}{1} + \dfrac{1}{a} + \dfrac{1}{b} + \dfrac{1}{c} + \cdots + \dfrac{1}{n} = 2$. We then multiply this equation by n to get $\dfrac{n}{1} + \dfrac{n}{a} + \dfrac{n}{b} + \dfrac{n}{c} + \cdots + \dfrac{n}{n} = 2n$, which is the same as $\dfrac{n}{a} + \dfrac{n}{b} + \dfrac{n}{c} + \cdots + \dfrac{n}{n} = n$, where the left side of the equation consists of the proper divisors of n in decreasing order. This is essentially the definition of a perfect number, which is one whose proper divisors have a sum equal to the number itself. We have noticed before that 6 is a perfect number, since the sum of the proper divisors $1 + 2 + 3 = 6$. The next five larger perfect numbers are: 28, 496, 8,128, 33,550,336, and 8,589,869,056.

When we take the next larger perfect number, 28, whose proper divisors are 1, 2, 4, 7, 14, we can do the following: $\dfrac{1}{1} + \dfrac{1}{2} + \dfrac{1}{4} + \dfrac{1}{7} + \dfrac{1}{14} + \dfrac{1}{28} = 2$. This procedure will work for any perfect number. However, for a larger number such as 496, the task would be overwhelming. Once again, the trickster will expose a unique feature of mathematics.

TRICK 51: THE SECRET AREA

In its first attempt, the audience will try to find the various line segments that are not yet fully marked and then the areas of the square and the isosceles right triangle. This is merely a distraction. The trick to solving this problem is to rotate the triangle about the right-angle vertex until the sides of the triangle contain the two vertices of the square, as shown in Figure 3.40. It can be easily shown that the two darker shaded triangles are congruent. So the area of the original quadrilateral is, therefore, equal to the area in the square formed by the two equal sides of the isosceles right triangle and the side of the square. That area is exactly $\frac{1}{4}$ of the area of the square. Therefore, the area of that original quadrilateral is $\frac{1}{4}$ of the area of the square, which is $\frac{1}{4} \cdot 81 = 20\frac{1}{4}$.

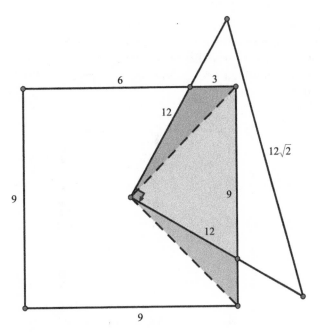

Figure 3.40

TRICK 52: REMOVING MATCHSTICKS TO MAKE FEWER SQUARES

It is relatively easy to remove two matchsticks, leaving three squares, as shown in Figure 3.41.

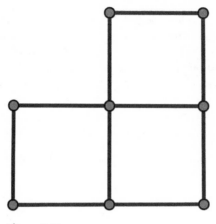

Figure 3.41

However, to remove two matchsticks, leaving only two squares, is a bit more challenging, as your audience will find out when presented with the task. This is shown in Figure 3.42. One might say that the result here is rather counterintuitive.

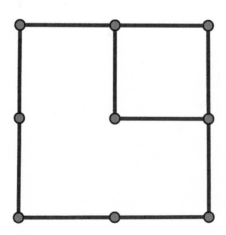

Figure 3.42

TRICK 53: MOVING MATCHSTICKS TO MAKE MORE SQUARES

Figure 3.43 shows the three matchsticks removed, forming five squares, namely the four small squares and the large overarching square. The solution here is quite unexpected and shows the power of logical thinking.

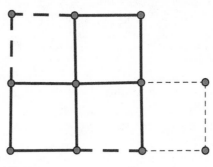

Figure 3.43

TRICK 54: REPRESENTING THE NUMBER ONE

The trickster could show this as a rather simple challenge, because the number 1 to any power is still equal to 1. This would allow him to respond to the original challenge with the following: $1^{2,3456,789}$. However, a more creative response to the challenge would be this one:

$1+(2-3)+(5-4)+(6-7)+(9-8)= 1+(-1)+(1)+(-1)+(1)=1$.

TRICK 55: AVERAGE SPEEDS

Here the trickster will upset the audience by telling them that 45 is a wrong answer. The reason why this is wrong answer is that the driver spent *twice as much* time driving at 30 mph as he did driving at 60 mph. Therefore, the two rates of speed cannot be given the same weight. The correct response would be to give the appropriate weights based on time spent, so that the correct average speed would be $\dfrac{30 + 30 + 60}{3} = 40$ mph.

For those not convinced by this argument, try something a bit closer to home. Pose a question about the grade a student deserves who scored 100% on 9 of 10 tests in a semester and on one test scored only 50%. Would it be fair to assume that this student's performance for the term was 75% (i.e., $\dfrac{100 + 50}{2}$)?

The reaction to this suggestion will tend toward applying appropriate weight to the two scores in consideration. The 100% was achieved nine times, and the 50% was achieved only once. Therefore, 100% ought to get the appropriate weight. Thus, a proper calculation of the student's average ought to be $\dfrac{9(100) + 50}{10} = 95$. This clearly appears to be more just!

An astute reader may now ask, "What happens if the rates to be averaged are not multiples of one another?" For the speed problem above, one could find the time "going" and the time "returning" to get the total time, and then with the total distance calculate the total rate, which is, in fact, the average rate.

There is a more efficient way to consider this problem, which enables us to consider some uncommon concepts in mathematics. We introduce a concept called the *harmonic mean*. This is the mean of a harmonic sequence, a sequence of numbers whose reciprocals form an arithmetic sequence, that is, one with common differences. The name harmonic may come from the fact that one such sequence is $\dfrac{1}{2}, \dfrac{1}{3}, \dfrac{1}{4}, \dfrac{1}{5}, \dfrac{1}{6}, \dfrac{1}{7}, \dfrac{1}{8}$. Guitar strings of these relative lengths when strummed together produce a harmonious sound.

This frequently misunderstood mean (or average) usually causes confusion. But once we discern that we are to find the average of rates (i.e., the harmonic mean), we have a useful formula for calculating the harmonic mean for rates over the same base. In the above situation, the rates were for the same distance (round-trip legs).

The harmonic mean for two rates, a and b, is $\dfrac{2ab}{a + b}$,

and for three rates, a, b, and c, the harmonic mean is $\dfrac{3abc}{ab + bc + ac}$.

You can see the pattern evolving, so that for four rates the harmonic mean is $\dfrac{4abcd}{abc + abd + acd + bcd}$.

Now applying the harmonic mean formula for two rates to the above speed problem gives us $\dfrac{2 \times 30 \times 60}{30 + 60} = \dfrac{3600}{90} = 40$.

This topic is not only useful, but also serves to sensitize the audience to the notion of weighted averages, a very important concept to remember.

TRICK 56: CREATING A FAMOUS MAGIC SQUARE

Now let's see how to construct a 4 × 4 magic square. We begin with a sequential square arrangement of numbers, as shown in Figure 3.44, and replace all the numbers in the diagonals with the number that, when added to them, yields 17.

This produces the square arrangement of numbers shown in Figure 3.45. This is a magic square, since all the rows, columns, and diagonals add up to 34.

At this point we can say the trickster has done what was expected to create a magic square. However, the square in Melencolia I has the two center columns interchanged, resulting in the magic square shown in Figure 3.46. This allowed Dürer to use the two center cells in the bottom row to indicate the year (1514) in which this etching was done.

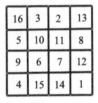

Figure 3.44   **Figure 3.45**   **Figure 3.46**

First let's make sure that it is a magic square. The sums of all the rows and all the columns must be equal, with all being 34. That is all that would be required to consider this square matrix of numbers a "magic square." However, this "Dürer Magic Square" has lots more properties that other magic squares do not have. This should give the trickster even more material to impress the audience.

- The four corner numbers have a sum of 34.

$$16 + 13 + 1 + 4 = 34$$

- Each of the four corner 2 × 2 squares has a sum of 34.

$$16 + 3 + 5 + 10 = 34$$
$$2 + 13 + 11 + 8 = 34$$
$$9 + 6 + 4 + 15 = 34$$
$$7 + 12 + 14 + 1 = 34$$

- The center 2×2 square has a sum of 34.

$$10 + 11 + 6 + 7 = 34$$

- The sum of the numbers in the diagonal cells equals the sum of the numbers in the cells not in the diagonal.

$$16 + 10 + 7 + 1 + 4 + 6 + 11 + 13 = 3 + 2 + 8 + 12 + 14 + 15 + 9 + 5 = 68$$

The sum of the squares of the numbers in the diagonal cells equals the sum of the squares of the numbers not in the diagonal cells.

$$16^2 + 10^2 + 7^2 + 1^2 + 4^2 + 6^2 + 11^2 + 13^2 = 3^2 + 2^2 + 8^2 + 12^2 + 14^2 + 15^2 + 9^2 + 5^2 = 748$$

- The sum of the cubes of the numbers in the diagonal cells equals the sum of the cubes of the numbers not in the diagonal cells.

$$16^3 + 10^3 + 7^3 + 1^3 + 4^3 + 6^3 + 11^3 + 13^3 = 3^3 + 2^3 + 8^3 + 12^3 + 14^3 + 15^3 + 9^3 + 5^3 = 9248$$

- The sum of the squares of the numbers in both diagonal cells equals the sum of the squares of the numbers in the first and third rows.

$$16^2 + 10^2 + 7^2 + 1^2 + 4^2 + 6^2 + 11^2 + 13^2$$
$$= 16^2 + 3^2 + 2^2 + 13^2 + 9^2 + 6^2 + 7^2 + 12^2 = 748$$

- The sum of the squares of the numbers in both diagonal cells equals the sum of the squares of the numbers in the second and fourth rows.

$$16^2 + 10^2 + 7^2 + 1^2 + 4^2 + 6^2 + 11^2 + 13^2$$
$$= 5^2 + 10^2 + 11^2 + 8^2 + 4^2 + 15^2 + 14^2 + 1^2 = 748$$

- The sum of the squares of the numbers in both diagonal cells equals the sum of the squares of the numbers in the first and third columns.

$$16^2 + 10^2 + 7^2 + 1^2 + 4^2 + 6^2 + 11^2 + 13^2$$
$$= 16^2 + 5^2 + 9^2 + 4^2 + 2^2 + 11^2 + 7^2 + 14^2 = 748$$

- The sum of the squares of the numbers in both diagonal cells equals the sum of the squares of the numbers in the second and fourth columns.

$$16^2 + 10^2 + 7^2 + 1^2 + 4^2 + 6^2 + 11^2 + 13^2$$
$$= 3^2 + 10^2 + 6^2 + 15^2 + 13^2 + 8^2 + 12^2 + 1^2 = 748$$

- Notice the following beautiful symmetries:

$$2 + 8 + 9 + 15 = 3 + 5 + 12 + 14 = 34$$
$$2^2 + 8^2 + 9^2 + 15^2 = 3^2 + 5^2 + 12^2 + 14^2 = 374$$
$$2^3 + 8^3 + 9^3 + 15^3 = 3^3 + 5^3 + 12^3 + 14^3 = 4624$$

- The sum of each adjacent upper and lower pair of numbers (vertically) produces a pleasing symmetry:

$16 + 5 = 21$	$3 + 10 = 13$	$2 + 11 = 13$	$13 + 8 = 21$
$9 + 4 = 13$	$6 + 15 = 21$	$7 + 14 + 21$	$12 + 1 = 13$

- The sum of each adjacent upper and lower pair of numbers (horizontally) likewise produces a pleasing symmetry:

$16 + 3 = 19$	$2 + 13 = 15$
$5 + 10 = 15$	$11 + 8 = 19$
$9 + 6 = 15$	$7 + 12 = 19$
$4 + 15 = 19$	$14 + 1 = 15$

Can you find some other patterns in this beautiful magic square? Remember, this is not a typical magic square, which would simply require that all the rows and columns have the same sum. The Dürer magic square has many more properties, which should give the trickster even more material to entertain and impress the audience.

TRICK 57: CONSTRUCTING A BASIC ODD-ORDER MAGIC SQUARE

Here the trickster can show the trick for making an odd-order magic square. There are many places where we can begin placing the numbers in numerical order from 1 to 9. We shall begin by placing the number 1 in the middle cell of the top row. We will place successive numbers above and to the right. The number 2 needs to be placed diagonally above and to the right, which is off the square. So we place it at the bottom of the row, where it would have landed on the square had it been larger, as shown in Figure 3.47.

The next number, 3, is then placed above and to the right of the 2. Once again, that goes off the grid, so we place it at the beginning of the row in which it landed, as shown in Figure 3.48.

The number 4 would normally go above and to the right of the number 3. But since that cell is blocked, we simply place the number 4 below the number 3 and then proceed diagonally upwards until we get to the number 6, as shown in Figure 3.49.

Our next move, shown in Figure 3.50, would be to place the 7 above and to the right. However, there is no square to accommodate that move, so it is placed below the 6. And we then continue that pattern, placing the 8 above and to the right of the 7, which requires moving the 8 to the far left of the first row. And once again the 9 would be above and to the right of the 8, but it is off the square and therefore goes to the bottom of that row. This completes the magic square, which we can now test to see if each row and each column and the diagonals have the same sum, which here is 15.

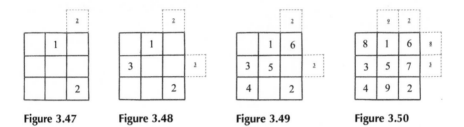

Figure 3.47   Figure 3.48   Figure 3.49   Figure 3.50

TRICK 58: CONSTRUCTING OTHER ODD-ORDER MAGIC SQUARES

The procedure to complete a 5 × 5 magic square is essentially the same as that for the previous 3 × 3 magic square. Looking at the square in Figure 3.51, begin at the middle cell of the top row and always move up and to the right. When a cell is off the square, you place the number at the end of the row or column. Once again, we have created a magic square, where each of the rows, columns, and diagonals has a sum of 65. The trickster should guide the audience so that they can then construct even larger odd-order magic squares and impress others!

18	25	2	9		
17	24	1	8	15	17
23	5	7	14	16	23
4	6	13	20	22	4
10	12	19	21	3	10
11	18	25	2	9	

Figure 3.51

TRICK 59: CREATING THE RADIUS

This problem appears to be rather complicated, since the result seems counter-intuitive. Remember, as shown in Figure 3.52, the point P is *any* point along the diameter. Yet the line segment AB will be shown to be the same length as the radius.

To show that AB is equal in length to the radius of the semicircle, we begin by constructing auxiliary lines, as shown in Figure 3.53. Since parallel lines cut off equal arcs along a circle, we can conclude—from the parallel lines drawn in the figure through point M, the center of the circle—that the arcs AX, BY, and CZ are equal. Therefore, we have equal arcs AB and XY, as their respective chords AB and XY are equal. However, the chord XY is the third side of an isosceles triangle whose vertex angle $\angle XMY = 60°$. That makes the triangle XMY equilateral. Thus, we have shown that the chord XY is equal to the radius of the circle, since it is equal to XM, and it follows that AB, which is equal in length to XM, is equal to the radius of the circle as well.

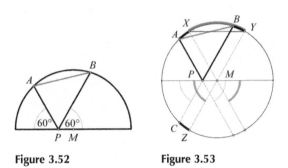

Figure 3.52   **Figure 3.53**

TRICK 60: BLACK AND RED CARDS IN A DECK

Typically, the trickster can approach this problem by representing the number of black cards and the number of red cards in each pile in a symbolic fashion. We can represent the situation symbolically as follows:

B_1 = the number of black cards in pile #1
B_2 = the number of black cards in pile #2
R_1 = the number of red cards in pile #1
R_2 = the number of red cards in pile #2

Then, since the total number of black cards equals 26, we can write this as $B_1 + B_2 = 26$, and since the total number of cards in pile #2 equals 26, we have $R_2 + B_2 = 26$.

By subtracting these two equations: $B_1 + B_2 = 26$, and $R_2 + B_2 = 26$, we get: $B_1 - R_2 = 0$. Therefore, we have $B_1 = R_2$, which tells us that the number of red cards in one pile equals the number of black cards in the other pile. Although this solves the problem, there is nothing particularly elegant about the solution.

As an alternate and perhaps a cleverer approach, we shall take all the red cards in pile #1 and switch them with the black cards in pile #2. Now, all the black cards will be in pile #1, and the red cards in pile #2. Therefore, the number of red cards in one pile and the number of black cards in the other pile had to be equal to begin with. Simple logic solves the problem, and the trickster has another impressive showing!

NOTE

1. Books recommended are *New Recreations with Magic Squares*, by W. H. Benson and O. Jacoby (New York: Dover, 1976) and *Magic Squares and Cubes*, by W. S. Andrews (New York: Dover, 1960). A concise treatment can be found in *Teaching Secondary School Mathematics: Techniques and Enrichment Units*, 10th edition, by A. S. Posamentier and B. S. Smith (World Scientific Publishing, 2021), 438–43. Also *Numbers: Their Tales, Types, and Treasures*, by A. S. Posamentier and B. Thaller (Prometheus Books, 2015).
2. *Strategy Games to Enhance Problem Solving Ability in Mathematics*, by A.S. Posamentier and S. Krulik (World Scientific Publishing, 2017).

4

Probability's Astonishing Conundrums

The field of probability provides us with a somewhat different arena for doing tricks. Here a trick typically presents the audience with a situation where the outcome is quite unexpected and sometimes very difficult to accept as correct. There are countless unusual—and often incomprehensible—examples that can clearly trick an audience, but can be proved correct to the surprise of the audience. We focus here on examples that can be justified with simple probability techniques. Some of these examples can be found in greater detail in other books, such as *Mathematical Amazements and Surprises* (Prometheus, 2009). Others are listed in the reference section of this book. We begin with one of the most surprising examples that illustrates the unusual aspects of probability, and show how it can be quite counterintuitive.

THE SURPRISING TRICK OF BIRTHDAY MATCHES

Here we present a most unexpected trick in mathematics. It is an excellent way to convince the uninitiated of the "power" of probability. Aside from being entertaining, it may upset your audience's sense of intuition.

Let us suppose you are in a room with about 34 people. What do you think the chances (or probability) are of at least 2 of these people having the same birth date (month and day, only)? Intuitively, one usually thinks about the likelihood of 2 people having the same date out of a selection of 365 days (assuming no leap year). Translating into mathematical language: 2 out of 365 would be a probability of $\dfrac{2}{365} = .005479 \approx \dfrac{1}{2}\%$, which represents a rather minuscule chance. With this trick you can actually demonstrate some very surprising results, and the audience will truly be awed.

To begin, let's consider a well-known group of 35 people, who were not selected on the basis of their birthdates, such as the first 35 presidents of the United States. Curiously, there are two presidents with the same birth date: the 11th president, James K. Polk (born on November 2, 1795), and the 29th president, Warren G. Harding (born on November 2, 1865).

Figure 4.1
James K. Polk

Figure 4.2
Warren G. Harding

In fact, for a group of randomly selected 35 people, the probability that at least 2 members will have the same birth date is greater than 8 out of 10, or $\frac{8}{10} = 80\%$. That is to say, if you had 10 groups, each having about 35 people, chances are very high that 8 of these 10 groups will have a birthday match.

For groups of 30 people, the probability that there will be a match of birthdates is a bit greater than 7 out of 10; that is, in about 7 of 10 randomly selected groups there would be a match of birthdates. What causes this incredible and unanticipated result? Can it be true? It seems to go against our intuition. Here the trickster can even make a "show" of this unexpected result. Perhaps more importantly, its explanation will give the audience some genuine insight into probability and what it means.

The audience will probably ask how this probability can be so high when there are 365 possible birthdates?[1] Let us consider the situation in detail and the reasoning that will convince you that these probabilities are true and will substantiate your trick. Consider a class of 35 students. What do you think the probability is that one selected student matches his own birth date? Clearly, that is *certainty*, or 1. This can be written as $\frac{365}{365}$. The probability that another student does *not* match the first student (i.e., has a different birth date) is:

$$\frac{365 - 1}{365} = \frac{364}{365}.$$

The probability that a third student does *not* match the first and second students is:

$$\frac{365 - 2}{365} = \frac{363}{365}.$$

The probability of all 35 students *not* having the same birth date is the *product* of these probabilities:

$$p = \frac{365}{365} \cdot \frac{365 - 1}{365} \cdot \frac{365 - 2}{365} \cdots \frac{365 - 34}{365}$$

Since the probability (q) that at least 2 students in the group *have* the same birth date and the probability (p) that *no* 2 students in the group have the same birth date is a certainty (i.e., there is no other possibility), the sum of those probabilities must be 1, which represents certainty.

Thus, $p + q = 1$, and it follows that $q = 1 - p$.

In this case, by substituting for p we get:

$$q = 1 - \frac{365}{365} \cdot \frac{365 - 1}{365} \cdot \frac{365 - 2}{365} \cdots \frac{365 - 34}{365} \approx 0.8143832388747152.$$

In other words, the probability that there will be a birthdate match in a randomly selected group of 35 people is somewhat greater than $\frac{8}{10}$. This, at first glance, is quite unexpected when one considers that there were 365 dates from which to choose. The motivated reader may want to investigate the nature of

Number of people in group	Probability of a birth date match	Probability (in percent) of a birth date match
10	.1169481777110776	11.69 %
15	.2529013197636863	25.29 %
20	.4114383835805799	41.14 %
25	.5686997039694639	56.87 %
30	.7063162427192686	70.63 %
35	.8143832388747152	81.44 %
40	.891231809817949	89.12 %
45	.9409758994657749	94.10 %
50	.9703735795779884	97.04 %
55	.9862622888164461	98.63 %
60	.994122660865348	99.41 %
65	.9976831073124921	99.77 %
70	.9991595759651571	99.92 %

Figure 4.3

this probability function. Figure 4.3 provides a few values to serve as a guide, as you may want to be prepared to present this trick.

Notice how quickly "almost-certainty" is reached—very close to 100%. With about 60 people in a room, the chart indicates that it is almost certain (99% probability) that 2 people will have the same birthdate.

Just as an aside, were you to do this with the death dates, for example, of the first 35 presidents, you would notice that two died on March 8 (Millard Fillmore in 1874 and William H. Taft in 1930) and *three* presidents died on July 4 (John Adams and Thomas Jefferson in 1826, and James Monroe in 1831). The latter case would give the impression that a person can, to some extent, will their death date.

From the table figure 4.3, we see that in a group of 30 people the probability of there being two people with the same birth date is about 70.63%. We need to caution the trickster *not* to go into a room with 30 people and say that there is a 70% chance that one person in the room will share your birthdate. The probability of having someone in the room with your birthdate is about 7.9%—considerably lower, since we now seek a specific birthdate rather than just a match of any birthdate. Let's see how we can determine this probability.

We will find the probability that there are no matches to your birthdate and then subtract that probability from 1, which is certainty.

$$p_{\text{Probability that no one has your birthday}} = \left(\frac{364}{365}\right)^{30}$$

The probability that one of these 30 people has my birth date is then:

$$q = 1 - p_{\text{Probability that no one has your birthday}} = 1 - \left(\frac{364}{365}\right)^{30} \approx 0.079008598089550769.$$

As you go about entertaining your audience with an extension of this trick you might mention that in a randomly selected group of 253 people in a room, the probability of one person having the same birthday as one specific person in the audience is about 50%.[2] If nothing else, this astonishing demonstration should be an eye-opener about the inadvisability of relying too much on intuition. Not only will you have provided a trick for others to use in various contexts, but you will also have shed light on the power of probability.

PROBABILITY IN GEOMETRY

Here is a trick that you will need to explain after you have demonstrated it. Begin by preparing in advance two concentric circles, where the radius of the smaller circle is one-half the radius of the larger circle, as shown in Figure 4.4.

What is the probability that a point selected in the larger circle is also in the smaller one? The typical (and correct) answer is $\frac{1}{4}$. This can easily be shown by letting the smaller circle's radius, OA, be represented by r, and the large circle's radius be represented by R, where $r = \frac{1}{2}R$. Then the area of

the small circle is $\pi\left(\frac{1}{2}R\right)^2 = \frac{1}{4}\pi R^2$, that is, $\frac{1}{4}$ of the area of the larger circle, which is πR^2.

Therefore, if a point is selected at random in the larger circle, the probability that it would be in the smaller circle as well is $\frac{1}{4}$.

Now here is where the trick comes in, as you ask your audience to look at this question differently. The randomly selected point P must lie on some

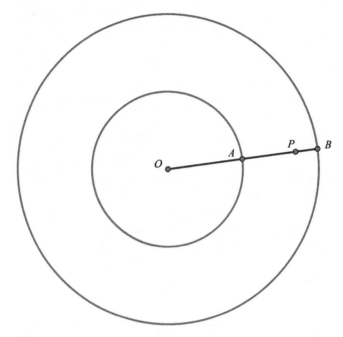

Figure 4.4

radius line of the larger circle. Let's say we use radius OAB, where A is its midpoint. The probability that a point P on OAB would be on OA (i.e., in the smaller circle) is $\frac{1}{2}$, since $OA = \frac{1}{2}OB$. Now, if we were to do this for any other point in the larger circle, the probability of the point being on a radius line and in the smaller circle would be $\frac{1}{2}$. This, of course, is *not* correct, although it seems perfectly logical. Where is the error? Here is where you expose the trick, which is actually an attempt to explain a conundrum. The "error" lies in the initial definition of each of two different sample spaces, that is, the set of possible outcomes of an experiment. In the first case, the sample space is the entire area of the larger circle, while in the second case, the sample space is the set of points on a radius such as OAB. Clearly, when a point is selected on OAB, the probability that the point will be on OA is $\frac{1}{2}$. These are two entirely different problems, even though (to dramatize the issue) they appear to be the same. Conditional probability is an important concept to stress.

And what better way to instill this idea than through a demonstration that shows obvious absurdities? Perhaps this conundrum will motivate the audience to investigate this topic further.

CREATING FAVORABLE ODDS

There are times when a trickster might want to show the audience how they can stay ahead of the game in a rather simple setting. In this game a couple is called to the stage, where there are 4 marbles, 2 of them white and 2 of them black. Also available are 2 cups. One member of the couple is asked to place the marbles into the 2 cups in the most favorable way for the other member to select a black marble in one try from one of the cups without looking.

The typical response would be to use both cups and not to leave 1 cup empty since that would immediately determine a loser—choosing that empty cup.

Now the question is, should each cup be filled equally with a black and a white marble? (See Figure 4.5.)

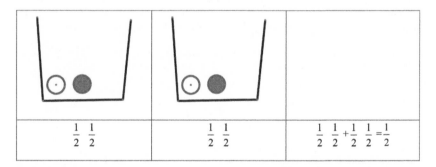

Figure 4.5

If that is the case, there is a $\dfrac{1}{2}$ chance of selecting one of the cups and a $\dfrac{1}{2}$ chance of selecting a black marble. So, for each cup there is a $\dfrac{1}{2} \times \dfrac{1}{2} = \dfrac{1}{4}$ chance of getting the black marble. Since that is the probability for each cup, we add $\dfrac{1}{4} + \dfrac{1}{4} = \dfrac{1}{2}$ to get the probability of getting a black marble under these circumstances, as shown in Figure 4.5.

On the other hand, if we place a black marble in one cup and the remaining marbles in the second cup, then the probability will become more

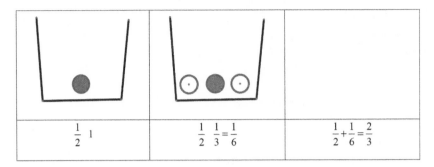

Figure 4.6

favorable, as shown in Figure 4.6. The probability of selecting the first cup containing one black marble is $\frac{1}{2}$, and then it is a certainty that the black marble will have been selected, with a probability of 1. The probability of selecting the second cup containing 2 black marbles and 1 white marble is again $\frac{1}{2}$, but this time the probability of selecting a black marble within this cup is $\frac{1}{3}$, as shown in Figure 4.6. Therefore, the probability of selecting a black marble is $\frac{1}{2} + \frac{1}{6} = \frac{2}{3}$, which is then the more favorable situation. This is a trick that can be used to astound the audience.

MAKING THE RIGHT CHOICE

Here is a simple probability question that the audience will appreciate. It will also enable them to better understand probability. Figure 4.7 shows 2 spinners, each partitioned into 3 colors. Your audience is asked to determine which spinners they would like to use to maximize their chances of getting a purple color paint, which is a mixture of red paint and blue paint. They have the opportunity for 2 spins. They can use either spinner number 1 or spinner number 2, once each or one spinner twice.

After letting your audience determine what they believe is the best chance for getting a purple color paint (that is, a mixture of red and blue paint), you might enlighten them with the following probability reasoning. Here the trickster can present an answer that will truly surprise most participants.

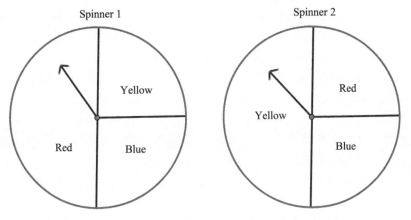

Figure 4.7

In order to get purple, we need to mix red and blue paint. There are 4 ways to choose the spinners, but 2 of them produce identical results.

We can use spinner 1, then use spinner 1 again.
We can use spinner 2, then spinner 2 again.
Or we can use spinner 1 and spinner 2 in any order.
Let's see how these options compare.

If we **use spinner 1 twice**, there are two ways to get purple:

On the first spin, there is a $\frac{1}{2}$ chance that we will get red, then a $\frac{1}{4}$ chance that we will get blue on the next spin, which offers us a $\frac{1}{2} \cdot \frac{1}{4} = \frac{1}{8}$ chance of getting purple.

On the first spin with spinner number 1, there is also a $\frac{1}{4}$ chance that we will get blue. Then on the next spin, there is a $\frac{1}{2}$ chance that we will get red, for a $\frac{1}{2} \times \frac{1}{4} = \frac{1}{8}$ chance of getting purple.

Adding these together tells us that there is a $\frac{1}{8} + \frac{1}{8} = \frac{1}{4}$ chance we will get purple by spinning spinner 1 twice.

Now let us consider **using spinner 2 twice**. Then we can have the following:

The probability of getting a red and then a blue is $\frac{1}{4} \times \frac{1}{4} = \frac{1}{16}$.

Or similarly, using spinner number 2 again, we can get blue and then red for a probability of $\frac{1}{4} \times \frac{1}{4} = \frac{1}{16}$. These add to give a probability of

$$\frac{1}{16} + \frac{1}{16} = \frac{1}{8}.$$

Finally, we can **spin both spinners 1 and 2**. The products yield the same result no matter which order we choose, so for this demonstration we will spin spinner number 1 first.

We have a $\frac{1}{2}$ chance of getting red, then a $\frac{1}{4}$ chance of getting blue. This yields a probability of getting purple of $\frac{1}{2} \times \frac{1}{4} = \frac{1}{8}$.

Or when using spinner 2 we have a $\frac{1}{4}$ chance of blue, then a $\frac{1}{4}$ chance of getting red $\frac{1}{4} \times \frac{1}{4} = \frac{1}{16}$. When we add these together, we get a probability of $\frac{1}{8} + \frac{1}{16} = \frac{3}{16}$, leaving the best probability, that of $\frac{1}{4}$. Therefore, the best option is to spin spinner number 1 twice. This is not only a trick that can be used in various ways, but also a wonderful demonstration of how probabilities are computed.

THE MONTY HALL PROBLEM ("LET'S MAKE A DEAL")

Here we merely recall a trick that was played on thousands of people watching the long-running television game show *Let's Make a Deal*. This trick has been the topic of entire books, since it presents a situation that is truly counterintuitive and controversial. The show featured a problematic situation for a randomly selected audience member, who came on stage and was presented with 3 doors. (See Figure 4.8.) Two of the 3 doors had donkeys behind them, and 1 concealed a car. The audience member was tasked with guessing which door concealed

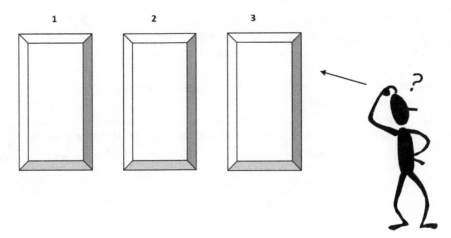

Figure 4.8

the car, which would be awarded as a prize if he selected it. There was only one wrinkle: after the contestant made an initial selection, the host, Monty Hall, exposed a donkey behind one of the 2 unselected doors (leaving 2 doors still unopened). The audience participant was then asked if he or she wanted to stay with his or her original selection (not yet revealed) or to switch to the other unopened door. At this point, to heighten the suspense, the rest of the program's audience would shout out "stay" or "switch" with seemingly equal frequency. The question is what to do? Does it make a difference? If so, which is the better strategy (i.e., the greater probability of winning) to use here?

Let us look at this now step-by-step. The result may gradually become clear. There are 2 donkeys and 1 car behind these 3 doors. (See Figure 4.9.) The

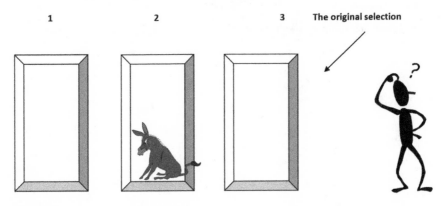

Figure 4.9

contestant must try to get the door with the car. Suppose that the contestant selects Door #3. Monty Hall opens one of the doors that the contestant *did not* select and intentionally exposes a donkey. (See Figure 4.9.)

He asks the contestant: "Do you still want your first-choice door, or do you want to switch to the other closed door?" This is where the great confusion and controversy begins. Most people feel that there is a 50-50 chance that it is behind one of the two closed doors. This question has also been argued by mathematicians for many years. To help make a decision, we shall consider an *extreme case*:

Suppose there were 1000 doors instead of just 3 doors. (See Figure 4.10.)

Figure 4.10

Let us further suppose that the contestant chooses Door #1000. How likely is it that the contestant chose the right door? ***"Very unlikely,"*** since the probability of getting the right door is $\dfrac{1}{1000}$. How likely is it that the car is behind one of the other doors? ***"Very likely"***: $\dfrac{999}{1000}$ (See Figure 4.11.)

Figure 4.11
These are all *"very likely"* doors

These are all *"**very likely**"* doors! (See Figure 4.12.)

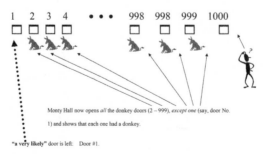

Monty Hall now opens *all* the donkey doors (2 – 999), *except one* (say, door No. 1) and shows that each one had a donkey.

"a very likely" door is left: Door #1.

Figure 4.12

We are now ready to answer the question: Which is a better choice?

- Door No. 1000 (*"very unlikely"* door), or
- Door No. 1 ("a very likely" door)?

The answer is now obvious. We ought to select the "very likely" door, which means "switching" is the better strategy for the audience contestant to follow. Using the extreme case provides a much easier way to see the best strategy than if we had tried to analyze the situation with only 3 doors. The principle is the same in either situation.

As mentioned earlier, this problem has caused many an argument in academic circles and was also a topic of discussion in *The New York Times* and other popular publications. John Tierney wrote in *The New York Times* (Sunday, July 21, 1991) that "perhaps it was only an illusion, but for a moment here it seemed that an end might be in sight to the debate raging among mathematicians, readers of *Parade* magazine and fans of the television game show 'Let's Make a Deal.' They began arguing last September (Sept. 9, 1990) after Marilyn vos Savant published a puzzle in *Parade*. As readers of her 'Ask Marilyn' column are reminded each week, Ms. vos Savant is listed in the Guinness Book of World Records Hall of Fame for 'Highest I.Q.,' but that credential did not impress the public when she answered this question from a reader." She gave the right answer, but still many mathematicians argued. Even the Hungarian-American mathematician Paul Erdös (1913–1996), considered one of the most prolific mathematicians of the twentieth century, was stumped by this problem and had to be convinced about the error of his thinking.[3]

THE THREE-CARD CHALLENGE

As a follow-up, a trickster might want to firm up the thinking presented with this trick. Suppose you have 3 cards: one has 2 sides blue, one has 2 sides red, and the third card has 1 side red and 1 side blue. Without looking at the bottom side of one of the 3 cards, it is placed on a table. If the side face up (showing) is blue, then clearly this card is not the red-red card. If it is the blue-blue card, then the bottom (face down) side is blue. If the card is the blue-red card, then the face-down side is red. You might be tempted to say that it is equally likely that the face-down side is red or blue. Well, by now you probably realize that this is not the case. This is analogous to the decision with the 2 closed doors mentioned in the previously—discussed Monty Hall problem. The correct analysis must take into account that we might have been looking at side 1 of the blue-blue card, or side 2 of the blue-blue card, or the blue side of the blue-red card. There are 3 possible blue sides—all equally likely. Two of them will lead to a face-down blue side and one to a face-down red. Therefore, the red and blue are not equally likely. Presenting this "tricky" thinking should help concretize the logic used earlier in the Monty Hall problem.

SHARPENING PROBABILITY THINKING

Tricks in mathematics sometimes show something that appears to be logical and is then proven to be wrong—that is, tricky! Such a situation can be shown by tossing a coin. Follow along as we consider the results of a coin being tossed and the numbers of heads and tails that appear.

Coin tossing case 1:
Suppose we toss 4 coins. What is the probability that there will be at least 2 heads? A description of the possible outcomes will bring some clarity to the question (H = heads; T = tails), as shown in Figure 4.13.

Number of heads	Possible outcomes by order of toss
4	HHHH
3	HHHT, HHTH, HTHH, THHH
2	HHTT, HTHT, TTHH, THHT, HTTH, THTH
1	TTTH, TTHT, THTT, HTTT
0	TTTT

Figure 4.13

There are 11 tosses that resulted in at least 2 heads out of 16 possible tosses, as shown in this same figure. This means that the probability of getting at least 2 heads is $\frac{11}{16} = 0.6875$.

Coin tossing case 2:
This time we will toss a coin 10 times. The first time, we had the following result:

1. H, T, T, H, T, T, T, H, T, H

The second time we tossed the coin 10 times, we got the following result:

2. H, T, H, T, H, T, H, T, H, T

The third time we tossed the coin 10 times, the result was:

3. H, H, H, H, H, H, H, H, H, H

The question we are faced with here is: Which of the three tosses is most likely to occur? One is tempted to select outcome No. 2, since it is assumed that unordered results are most likely to occur. This assumption is false! A nicely ordered result (such as tosses Nos. 2 and 3) is just as likely to occur as one that is unordered—such as No. 1.

For each toss, the probability of a head and a tail is equally likely—namely a probability of $\frac{1}{2}$. Therefore, the probability of each of the three results is equally likely with a probability of $\left(\frac{1}{2}\right)^{10} = \frac{1}{1024} = 0.0009765625$. Now that you, the trickster, understand these unexpected results, you can manufacture quite a few entertaining opportunities for the audience. For example, if you toss a coin 5 times, tell your audience you just got 5 heads, and ask them what they expect the next toss to be, the typical response is "a head." From the above discussion, we know that that is not necessarily the case, as there is still a 50-50 chance that a tail would appear. So, bear in mind that having a series of tails does not guarantee a head. We leave it to the reader's creativity to produce some entertaining tricks based on this counterintuitive finding. These are some of the subtle differences that must be accounted for when we look at questions of probability.

INTRODUCTION OF A SAMPLE SPACE

This trick can be a bit instructional as well as entertaining and enlightening. When doing a problem involving probability, set up the sample space—a set that shows all the possibilities—to see what is actually taking place. This trick could place you in a game situation, where your audience's intuition could work against them. Unless you actually set up the sample space, you may not be able to resolve the inequity of the game, or trick, we are about to present.

Begin by placing 1 red chip and 2 black chips in an envelope. Here are the rules for the game the trickster is about to play with the audience:

1. Without looking, the trickster draws 2 chips from the envelope.
2. If the colors of the 2 chips are different, the trickster scores a point. If they are the same, the friend scores the point. Next it is the other player's turn.
3. After each draw, the chips are returned to the envelope and the envelope is shaken and prepared for the next draw.
4. The first player to score 5 points is the winner.

The question is to determine if the game is a fair one (that is, each player has an equal chance of gaining a point). After playing the game several times, you might conclude that the game is not fair. The trickster should win the game most of the time. What single chip would you add to make the game fair? Typically, one would suggest adding a second red chip to the envelope. It might be best to look at this problem with the help of setting up a sample space.

Situation 1: 1 red chip and 2 black chips

The possible draws would be:

$$R_1B_1 \quad R_1B_2 \quad \mathbf{B_1B_2}$$

Thus, the friend has only 1 out of 3 chances of scoring a point, for a $\frac{1}{3}$ probability. Therefore, the original game is unfair.

Situation 2: Adding 1 red chip, so that we have 2 red chips and 2 black chips

The possible draws would be:

$$R_1B_1 \quad R_1B_2 \quad \mathbf{R_1R_2}$$
$$R_2B_1 \quad R_2B_2 \quad \mathbf{B_1B_2}$$

Surprise! The friend has only 2 out of 6 chances of scoring a point, for a $\frac{1}{3}$ probability. The game is, once again, unfair.

Situation 3: Adding another black chip to the original set of chips, we get 1 red chip and 3 black chips

The possible draws would be:

$$R_1B_1 \quad R_1B_2 \quad R_1B_3$$
$$B_1B_2 \quad B_1B_3 \quad B_2B_3$$

This time the friend has 3 out of 6 chances of scoring a point, for a $\frac{1}{2}$ probability. The game is now fair.

The use of the sample space reveals that your intuition does not necessarily yield a correct resolution of the problem. This makes the concept of a sample space indispensable as the trickster sets up a game.

USING SAMPLE SPACES TO SOLVE TRICKY PROBABILITY PROBLEMS

You probably see how identifying a sample space can be useful. We now shall try to further convince you of its value by showing you another problem that defies intuition—and which can be useful for a bag of tricks. Consider the following problem:

Three cardboard cards are marked as follows:

Card 1: both sides VOCAL
Card 2: both sides INSTRUMENTAL
Card 3: one side VOCAL, the other side INSTRUMENTAL

A person has just three phonograph records. The first has vocal performances on both sides, the second has instrumental music on both sides, and the third has vocal performances on one side and instrumental music on the other side. A person, who is in a darkened room, plays one of these records. What is the probability that he or she hears a vocal performance?

To make this problem a bit more manageable we shall use some symbols:

Denote by $v_{1,1}$: side 1 of record 1
Denote by $v_{2,1}$: side 2 of record 1
Denote by $i_{1,2}$: side 1 of record 2
Denote by $i_{2,2}$: side 2 of record 2
Denote by $v_{1,3}$: side 1 of record 3
Denote by $i_{2,3}$: side 2 of record 3

The sample space for this problem situation consists of the six equally likely possible outcomes: $v_{1,1}, v_{2,1}, i_{1,2}, i_{2,2}, v_{1,3}, i_{2,3}$. Three of these, $v_{1,1}, v_{2,1}, v_{1,3}$, consist of vocal performances. Therefore, the probability of hearing a vocal performance is $\frac{3}{6} = \frac{1}{2}$. This is a reasonable solution—one that conforms with our intuition.

Under the same situation as above, this person, who is in a darkened room, puts on one of these records and hears a vocal performance. What is the probability that the other side of that same record is also vocal? The usual response is $\frac{1}{2}$, with the "reasoning" that one of the two vocal records has a vocal on the other side. This reasoning is wrong! The explanation that follows should be useful for a trickster.

The correct reasoning is that since there are six record sides equally likely to be played and the person in the problem is playing a vocal side, the sample space for this is limited to the set $\{v_{1,1}, v_{2,1}, v_{1,3}\}$. Two $\{v_{1,1}, v_{2,1}\}$ of the three elements of the sample space set represent a vocal on the other side. Therefore, the probability is $\frac{2}{3}$. This typically wrongly–solved problem demonstrates the usefulness of setting up a sample space. Compare this "conditional probability" $\left(\frac{2}{3}\right)$ with the probability $\left(\frac{1}{2}\right)$ of the first problem above. This would make for a very fine trick!

PROBABILITY THROUGH COUNTING

A well-prepared trickster should know that a thinking adjustment is required when first encountering probability. The simple illustration, such as pulling an ace out of a deck of 52 cards, requires little imagination. However, a problem such as the following does require some probability-thinking skills, which can fortify the trickster:

> *Two identical paper bags contain red and black checkers.*
> <u>*Bag A*</u> *contains 2 red and 3 black checkers*
> <u>*Bag B*</u> *contains 3 red and 4 black checkers*
> *One bag is chosen at random and a red checker is drawn from it. What is the probability that Bag A was chosen?*

As we can see in Figure 4.14, since the first *even* common multiple of 5 and 7 (the bags' checker contents) is 70, we shall use this as the number of trials for our hypothetical experiment. Since the bag was chosen at random, we can assume for our convenience that each bag was chosen 35 times. In that case:

> From <u>Bag A</u>: a red checker will be selected 2 out of 5 times.
>   So, for 35 trials a red checker will be drawn 14 times.
> From <u>Bag B</u>: a red checker will be selected 3 out of 7 times.
>   So, for 35 trials a red checker will be drawn 15 times.

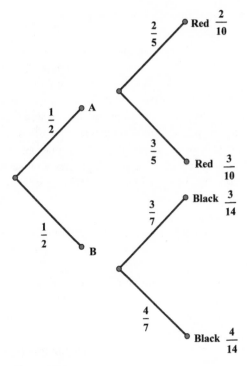

Figure 4.14

Therefore, a red checker will be drawn $14 + 15 = 29$ times, of which 14 would be from Bag A. Consequently, the probability that the red checker will be drawn from Bag A is $\dfrac{14}{9}$.

This displays the kind of comparison needed to succeed in determining the probability of an event.

THINK BEFORE COUNTING

Often a problem situation seems so simple that we plunge right in without first thinking about a strategy to use. This impetuous beginning often leads to a less elegant solution than one that results from a bit of forethought. Here are two examples of simple problems a trickster can present that can be made even simpler by thinking before working on them.

Find all pairs of prime numbers whose sum equals 999.

Some of your audience will begin by taking a list of prime numbers and trying various pairs to see if they obtain 999 for a sum. This is obviously very tedious as well as time consuming, and you would never be quite certain that you had considered all the prime number pairs. As trickster, we can use some logical reasoning to solve this problem. In order to obtain an odd sum (in this case 999) for two numbers (prime or otherwise), exactly one of the numbers must be even. There is only one *even* prime, 2. Therefore, there can be only one pair of primes whose sum is 999, and that pair is 2 and 997. The trickster scores again!

A second problem where preplanning or some orderly thinking makes sense is as follows:

A palindrome is a number that reads the same forwards and backwards, such as 747 or 1991. How many palindromes are there between 1 and 1000, inclusive?

The traditional approach to this problem would be to attempt to write out all the numbers between 1 and 1000, and then see which ones are palindromes. However, this is a cumbersome and time-consuming task at best, and some of them could easily be overlooked.

Let's see if we can look for a pattern to solve the problem in a more direct fashion.

Range	Number of Palindromes	Total Number
1–9	9	9
10–99	9	18
100–199	10	28
200–299 ·	10	38
300–399	10	48
400–899	10	108

There is a pattern. There are exactly 10 palindromes in each group of 100 numbers (after 99). Thus, there will be 9 sets of 10, or 90, plus the 18 from numbers 1 to 99, giving a total of 108 palindromes between 1 and 1000.

Another solution to this problem would involve organizing the data in a favorable way. Consider all the single-digit numbers (self-palindromes). There are nine such numbers. There are also nine two-digit palindromes. The three-digit palindromes have 9 possible "outside digits" and 10 possible "middle digits," so there are 90 of these. In total, there are 108 palindromes between 1 and 1000, inclusive. The motto the trickster should impart is: think first, then begin a solution!

COIN FLIPPING AND THE NUMBER LINE

Here is a trick that the trickster can play on his audience in any way he likes. You begin with a number line. For our purposes we will just use one from − 5 to +5, as shown in Figure 4.15.

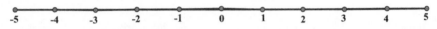

Figure 4.15

In this game a coin is placed at the 0 position. The coin is then flipped, and when it turns up a head (H), it is then moved one place to the right. When it turns up a tail (T), the coin is then moved one place to the left. At this point you can ask your audience to bet where the coin would be placed after five tosses. We will show that the probabilities of where the coin ends up are quite surprising. The chart shown in Figure 4.16 clearly indicates that the sum of the probabilities is equal to 1, which indicates certainty. In other words, the coin must land on one of these 11 positions. But your audience will be quite surprised to know that there are certain positions where the probability of the coin landing is 0. This will allow the trickster quite a few opportunities to "show off."

Landing position of the coin after five tosses	Probability of landing on this position
5	$\dfrac{1}{32}$
4	0
3	$\dfrac{5}{32}$
2	0
1	$\dfrac{10}{32}$
0	0
-1	$\dfrac{10}{32}$
-2	0
-3	$\dfrac{5}{32}$
-4	0
-5	$\dfrac{1}{32}$

Figure 4.16

A skilled trickster could extend the number line in both directions equally and also increase the number of coin tosses. However, the basic game is provided here.

THE WORTHLESS INCREASE

Here is a situation that you can use to play a trick on a friend. Present your friend with the following scenario: Suppose you have an item you want to sell him and are willing to give him a 10% discount if he buys it from you immediately. However, your friend decides that he would rather wait until the next day, at which point you tell him that you will have to raise the price 10%. Would your friend be correct in assuming that you are offering him the original price for the item before any discounts or increases? The answer is a resounding (and very surprising) no!

This little story is quite disconcerting. You would expect that with the same percent increase and decrease, you should be back where you started. This is intuitive thinking, but wrong! To explain this, you might have your friend choose a specific amount of money. Suppose we begin with $100. Calculate a 10% decrease and then a 10% increase. Using a $100 basis, we first calculate the 10% decrease to get $90, then the 10% increase, which is $9. This yields $99—$1 less than the beginning amount. Therefore, your friend benefits by having waited until the next day to buy your item.

You may wonder whether the result would have been different if we had first calculated the 10% increase on the $100 to get $110. Then we would take a 10% decrease of this $110, which is $11, to get $99—the same result as before. So, clearly the order makes no difference.

A similar situation, deceptively misleading, can be faced by a gambler. Offer your friend this challenge. You are presented with a chance to play a game where the rules are as follows: There are 100 cards, face down. Of those cards, 55 cards say "*win*" and 45 of the cards say "*lose*." You begin with $10,000. You must bet one-half of your money on each card turned over, and you either win or lose that amount based on what the card says. At the end of the game, all cards have been turned over. How much money do you have at the end of the game?

The same principle as above applies here. It is obvious that you will win ten times more than you will lose, so it appears that you will end with more than $10,000. But what is obvious is often wrong, and this is a good example. Let's say that you win on the first card; you now have $15,000. Then you lose on the second card; you now have $7500. If you had first lost and then won, you would still have $7500. So, every time you win one and lose one, you lose one-fourth of your money. So, you end up with $10,000 \cdot \left(\dfrac{3}{4}\right)^{45} \times \left(\dfrac{3}{2}\right)^{10}$. This is \$1.38 when rounded off. Surprised? With this knowledge, a mean trickster can surely take advantage of a "friend."

CALENDAR TRICKS

The calendar holds many surprising relationships that a trickster can use to entertain his audience.

Consider any calendar page, say, October 2002.

Sunday	Monday	Tuesday	Wednesday	Thursday	Friday	Saturday
		1	2	3	4	5
6	7	8	9	10	11	12
13	14	15	16	17	18	19
20	21	22	23	24	25	26
27	28	29	30	31		

Figure 4.17

Select a (3×3) square of any nine dates on the calendar, such as shown in Figure 4.17. Add 8 to the smallest number in the shaded region and then multiply that sum by 9: $(9 + 8) \times 9 = 153$. Then multiply the sum of the numbers of the middle row $(16 + 17 + 18 = 51)$ of this shaded matrix by 3. Surprise! It is the same as the previous answer, 153. But why? Here are some clues: the middle number is the mean (or average) of the 9 shaded numbers. The sum of the numbers in the middle column is $\dfrac{1}{3}$ of the sum of the 9 numbers. This gives you even more material with which to impress your audience.

While entertaining your audience with the calendar, have them consider what the probability is of 4/4[*], 6/6, 8/8, 10/10, and 12/12 all falling on the same day of the week. The "knee-jerk" reaction would be that it occurs about $\dfrac{1}{5}$ of the time. Wrong! The probability is 1, or certainty! But how can we justify this surprising result? Closer inspection will reveal that these dates are all exactly nine weeks apart. Such little-known facts always draw an interest that otherwise would be untapped. Counting on a calendar presents numerous surprises for a trickster to entertain an audience.

---

[*] 4/4 represents April 4, 6/6 represents June 6, and so on.

FLIPPING A LEGAL COIN

The trickster can sometimes entice the audience with the way a problem is phrased. Suppose Charlie flips a coin 5 times and it comes up heads 5 times in a row. What is the probability that he will obtain a head on the sixth flip?

Most people feel that this is highly unlikely; since Charlie had already gotten 5 heads, the next flip is most likely not going to be a head. The confusion lies in the fact that there are basically two kinds of questions that are often confused:

- What is the probability of obtaining a head on the sixth flip of a fair coin, when the first 5 flips have each already resulted in a head?
- What is the probability of obtaining 6 heads on 6 flips with a fair coin?

The question that the trickster proposed is actually the first one, where each flip is independently valued at a 50% chance of getting a head. Therefore, the answer is very simple: the chance of getting a head on the sixth flip is simply 50% or $\dfrac{1}{2}$.

However, it should be noted that the second question, that of getting 6 heads in a row, is much different. There the probability is $\left(\dfrac{1}{2}\right)^6 = \dfrac{1}{64}$

ESTABLISHING A CONSECUTIVE DIAGONAL

Sometimes a trick works because of its simplicity, which is overlooked by the audience. Consider the following question: How can you place the numbers 1 – 5 in a 5 × 5 square table of cells so that all the numbers 1 to 5 are in each column in any order you choose, and also represented in each of the diagonals? The arrangement you choose is the challenge.

The wording of the problem indicates that this is going to be somewhat of a challenge. The trickster should let the audience play with this a bit and then expose the simplicity of the solution. All the audience needs to do is list the numbers in each column sequentially from top to bottom, and likewise with each of the five numbers in the diagonals. You can see that the audience expected a rather tricky solution when, in fact, it was trivial, as shown in Figure 4.18.

1	1	1	1	1
2	2	2	2	2
3	3	3	3	3
4	4	4	4	4
5	5	5	5	5

Figure 4.18

A TRICK WITH A PAIR OF DICE

There is a dice game in which the trickster can take a strong advantage over his opponent. On the surface, the game seems fair; but the trickster knows that it is not and, therefore, chooses the advantageous position. Consider the following: two players, David and Lisa, compete by each tossing a regular six-sided die. The two numbers showing are then subtracted. If the difference is 0, 1, or 2, then David wins a point. If the difference is 3, 4, or 5, then Lisa wins a point. For example, if David rolls a 5 and Lisa rolls a 2, the difference is 3, which would give Lisa a point. After 15 rounds, the player with the most points is the winner of the match. Which player has the advantage in this game?

The trickster should know that when you consider all the differences, those that are 0, 1, and 2 have 24 ways of appearing, while those that are 3, 4, and 5 have only 12 ways of appearing. This can be easily seen by setting up a table of differences, as shown in Figure 4.19.

		David rolls					
		1	**2**	**3**	**4**	**5**	**6**
	1	**0**	1	2	3	4	5
	2	**1**	0	1	2	3	4
Lisa rolls	3	**2**	1	0	1	2	3
	4	**3**	2	1	0	1	2
	5	**4**	3	2	1	0	1
	6	**5**	4	3	2	1	0

Figure 4.19

An ambitious trickster may wish to change the rules from subtraction to addition and then determine the more advantageous position to take.

SOME TRICK QUESTIONS

Here are a few quick questions that have rather tricky responses.

- *Suppose there are three people at different parts of the Earth's globe. What is the probability that these three people are in the same hemisphere of the globe— regardless of any consideration of the poles of the earth.*

The answer to this question is usually unanticipated. Since any three points on a globe's surface will always be in the same hemisphere of the globe, regardless of its position, the probability is 1 or certainty.

- *A dart player is tossing 3 darts at a board that shows a 3 × 3 square with 9 cells. What is the probability that he will toss a win in tic-tac-toe?*

The number of combinations of 3 throws with 9 possible landings can be represented as $_9C_3 = \dfrac{9 \cdot 8 \cdot 7}{1 \cdot 2 \cdot 3} = 84$. However, only three rows, three columns, and two diagonals qualify for a win. Therefore, the probability is $\dfrac{3 + 3 + 2}{84} = \dfrac{2}{21}$.

- *Tell your audience that you have a drawer containing an odd number of black socks and an even number of blue socks. Ask them to determine the smallest number of black socks and blue socks that need to be in the drawer so that the probability of picking 2 black socks—without looking—is $\dfrac{1}{2}$.*

As trickster, you should know the answer is 15 black socks and 6 blue socks. It should be fun for the audience to determine the reason for this selection.

- *Ask your audience how many people would have to be in an auditorium before you can be sure of having 2 people with the same initials for their first and last names in that order.*

Here the trickster would have to show that for each of the 26 letters of the alphabet, there are 26 possibilities for the first name and another 26 possibilities for last name. To break it down a bit more clearly, if you were to make a listing, your first 26 possibilities would be: AA, AB, AC, AD, AE, ..., AX, AY, AZ. This would continue for each first name until one reaches the letter Z. Counting, this would be $26 \times 26 = 676$. Therefore, the audience would need to have at least 677 people to be certain that there are two with the exact same initials, first name and last name.

Tricks in the field of probability are often unpredictable because they seem to challenge our intuition. Yet through these tricks, the trickster can motivate the audience to pursue the concepts of probability—a subject often neglected in the school years—with some enthusiasm.

NOTES

1. Actually 366, but for the sake of our illustration—and simplicity—we will not address the possibility of birthdays on February 29.
2. For the calculation of this probability, see *Fifty Challenging Problems in Probability*, by Frederick Mosteller (Reading, MA: Addison-Wesley, 1965), 48–49.
3. Hoffman, Paul, *The Man Who Loved Only Numbers* (New York: Hyperion, 1998), 249–56.

Epilogue

You have gone through a plethora of mathematics tricks, intended not only to entertain your audience, but also to make them aware of the power and beauty of mathematics. Now you have the tools needed to demonstrate that mathematics can be fun beyond the prescribed school curriculum.

We hope that your audience will then pursue some of the topics covered in greater detail. They can be presented in a somewhat dramatic fashion—depending on the particular audience, whether a large group or an individual. It would be a great contribution to the common good to have motivated people to spread these ideas to others, thereby generating a multiplier effect that "infects" more people with a newfound love for mathematics. As the world becomes ever-more reliant on technology, the need to motivate an appreciation and, moreover, a love for mathematics becomes ever-so-much-more important, since mathematical thinking is key to success in information technology. It is hoped that this text's rather novel approach to fostering an enjoyment of mathematics will achieve this goal.